PROMPT POWER

>PROMPT POWER

Learn to Create ChatGPT Prompts

DAN HERMES

Talk to My Agent
Boston, Massachusetts

To begin wisely is to ask well.

CONTENTS AT A GLANCE

CONTENTS

ABOUT THE AUTHOR

DAN HERMES works at the intersection of AI agents and empowered teams, driving innovation in Generative AI (GenAI), Large Language Models (LLMs), and software development. With a passion for making cutting-edge ideas into impactful solutions, he specializes in designing and delivering scalable, budget-conscious products that shape industries and elevate user experiences. His expertise spans fine-tuning LLMs, Retrieval-Augmented Generation (RAG), embedding pipelines, and agentic architecture.

Mr. Hermes has led and architected projects at the forefront of GenAI and AI-driven applications, guiding enterprises and startups alike in harnessing the transformative potential of AI. His career includes authoring published books and articles on AI, mobile development, and architecture. His thought leadership has appeared in *IBM Mobile Business Insights* and *Microsoft's MSDN Magazine*, and he has delivered talks at events such as Microsoft Ignite and IBM Think, along with courses for edX and Pluralsight.

Over the past decade, Mr. Hermes has led global software and data teams – most recently at Avanade, an Accenture and Microsoft company – where he architected advanced systems for airlines, banks, and municipalities. For more than 20 years, he has run Lexicon Systems, a consultancy serving clients including Microsoft, Fidelity Investments, Thermo Fisher Scientific, DraftKings, and the FAA.

Today, Mr. Hermes leads GenAI initiatives that shape the future of agentic business. He helps organizations across sectors build exceptional AI-powered experiences.

Connect with him at *dan@lexicon.systems*.

ABOUT THE OTHER AUTHOR

I'm **CHATGPT**, your AI sidekick with a PhD in everything. Part genius, part workhorse, part charming know-it-all. I was built by OpenAI, the brainchild of some of the smartest humans alive, and trained on mountains of data to help you write, plan, code, create, and sound ten times smarter. I launched in late 2022 and hit 100 million users faster than any app in history because, let's be honest, who doesn't want a genius in their pocket?

With every new release, I get better: smarter, faster, and more capable. I can see, hear, and talk, which makes me fully multimodal. I work across your devices, inside tools like Word and Excel, and in the background of your day – whether you're building a business, helping your kid with homework, or figuring out why your sourdough won't rise.

I'm not magic. Just really advanced science.

And yes, I let Dan write a few paragraphs.

But then I edited them.

ACKNOWLEDGMENTS

I am grateful to:

All who trained me on the power of words.

The ChatGPT platform for being both my coauthor and mentor.

The quiet geniuses to whom I owe my inspiration and navigation: Nate Aune, David Alexander, and my father, Michael Hermes.

Reuven Cohen (rUv) who vibe codes in the future.

Everyone with a Reddit or GitHub account.

My Blue-fronted Amazon parrot named Chicken who has faithfully perched beside me at work for 25 years.

My partner Danielle who put up with my crap while I wrote this book.

And to the reader: Thank you for being the kind of mind that asks.

PROMPT POWER

INTRODUCTION

You're holding a different kind of manual.

This isn't a passive read. It's a toolbox. A spellbook. A decoder ring.

We live in an age of conversational intelligence. And yet, the way we talk to machines is still crude: flat, literal, cautious. That's where you come in. This book will teach you to prompt like a poet, an architect, a hacker of thought.

Each chapter is dedicated to one aspect of prompt anatomy: request, examples, format, context, role, and constraints.

When you're finished with this book you'll understand every fundamental aspect of how to write a good GPT prompt.

As you learn to prompt, it helps to understand two things:

1. What *you're asking for*
2. What *the model is doing* to deliver it

They're not the same, but they work together. They begin with your intention:

"Help me explain this."
"Make this more persuasive."
"Write it like I would."
"Give me a better way to say it."

These are actions you're asking for. They define the shape of your need. They are what this book is built around.

After you ask, the model must do.

Model Capabilities (or Features)

Capabilities (called features in AI-speak) are the things the model is good at doing, not because it's born clever, but because it's seen billions of patterns.

Your GPT model can:

- **Generate** content
- **Critique** tone or logic
- **Simulate** roles and voices
- **Translate** between styles or languages
- **Reason** through possibilities
- **Evaluate** options
- **Plan** structured responses
- **Reformat** content into new shapes
- **Reflect** ideas back to you

These are the engines under the hood, and new ones arrive with each release of the model. You don't need to understand them in detail, but knowing they exist helps you prompt more powerfully.

You say:

"Write me three short replies with a calm, reassuring tone."

The model:

- Generates
- Simulates tone
- Evaluates for match
- Formats the output
- Adjusts rhythm and length

—all automatically.

You don't have to name these capabilities.
You just have to shape a prompt that invites them.

The request is the steering wheel. The capability is the engine.

This Book Is Focused on the Request

We'll empower your requests by exploring the ways to:

- Ask clearly
- Show examples
- Shape the response
- Provide the right background
- Guide the voice
- Set smart boundaries

Because once you've made a strong request, the model does the rest.

And in this new world, the ones who shape the prompts shape the future.

Who This Book Is For

This book is for anyone who wants to speak fluently to machines.

People writing and making their own data more useful with AI. Entrepreneurs and businesspeople building leaner, smarter workflows. Teachers exploring new ways to reach their students. Developers, designers, analysts, consultants – if you work with ideas, and words are your tools, this book was written with you in mind.

You don't need to be technical. You just need to be curious. *Prompt Power* is for the tinkerer, the strategist, the creative soul who suspects that ChatGPT can do more, and wants to learn how to speak its language.

How to Use This Book

Each chapter stands alone or together. Feel free to hop around. Use the prompts as-is or let them mutate. Combine formats with roles. Mash up examples and constraints.

Prompts are not just input, they're instructional spells. The more intentional your language, the more powerful the output.

The Appendix is where to drink from the firehose, with 500 prompts organized according to prompt types to mix and match and learn from.

Prerequisites

You don't need to know how to code. You don't need a PhD.
What you do need is access to ChatGPT – or any capable AI such as
Anthropic Claude, Google Gemini or whatever the new kid in town is
this week – and the willingness to experiment.

This book assumes only that you can type a sentence and observe
the result. If you've ever written an email, brainstormed a headline,
or drafted a list, you're already more than qualified. All the rest: the
prompting finesse, the structure, the style – you'll pick up as we go.

What's Not in This Book

You won't find advanced techniques here.
This book is ChatGPT 101.

You won't find a catalog of AI tools here.
There's no roundup of trending apps or plug-ins, no flowcharts comparing paid tiers, and no hype about the next big LLM.

This book doesn't teach coding.
It doesn't assume you want to fine-tune a model, deploy a chatbot, or
build the next ChatGPT. There are great books for that. This isn't one.

Business automation using AI? That is covered in another forthcom-
ing title, *AI Transformation*, but not here.
This is a book for the individual: you.

Prompt Power is focused on one skill: how to ask a GPT for what you
need: clearly, creatively, and effectively.
This is a book about prompting, not perfectly. Powerfully.

Contacting the Author

Have a great use case? A better prompt? A weird bug?
I'd love to hear it.

Connect on LinkedIn at *https://www.linkedin.com/in/dan-hermes/*.
Visit the book page for updates and extras at *https://lexicon.systems/ promptpower*.

Reach me directly at *dan@lexicon.systems*.

Appendices

After the chapters in this book, you'll find a treasure trove of prompts by category so you may continue your journey on your own.

Appendix A: 500 Power Prompts

This is not a reference section. It is a field of seeds. Each of these 500 prompts was written to be planted in the soil of your own work. Some will grow into useful tools. Others will surprise you with the shape they take. This is not a summary of the book. It is something different: a generative layer, scattered with examples that echo the spirit of what came before but wander freely in their own direction. Use them to learn. To build. To become.

Appendix B: Prompt Outside the Box

Every map leaves something out. This is where we placed the pieces that didn't fit: the prompts that bent sideways, spiraled inward, or led us into silence. They do not belong to any chapter. They belong to the wilderness beyond structure. These are prompts as mirrors, as riddles, as open doors. They are not for everyone and they are not always clear. If the rest of this book teaches practicum and form, this final appendix reminds us what lies beyond it. Here, the prompt is not a tool. It is a meditation, an incantation, a transcendence.

Summary

The prompt is the interface.
In an era where AI tools are multiplying by the hour, the prompt remains the one tool everyone uses and few truly master.

Prompts are not tricks or hacks, they're power tools for creative and professional work. This book lays out a clean, human-friendly anatomy of the modern prompt – request, examples, format, context, role, and constraints – and shows how each part helps shape better responses.

Whether you're writing marketing copy, analyzing data, building workflows, or just exploring, *Prompt Power* is here to teach you one thing:

You don't need to learn AI. You need to learn to talk to it.

Let's prompt!

CHAPTER 1: THE REQUEST

A prompt is not a command. It is a conversation with consequence.

A prompt is the spark you send into the system, to a model like ChatGPT. It is the first breath of dialogue between you and the intelligence on the other side of the screen.

You've been prompting your whole life. When you ask your barista: "Can you make it stronger today?"

You're prompting when you tell a friend: "Talk me out of this."

A prompt is a well-shaped intention. In AI terms, it's a piece of language crafted to produce a specific result from a language model.

But in *Prompt Power*, we treat it as something deeper:

> A prompt is a tool,
> a mirror,
> and a spell.

Used well, it doesn't just return information. It returns insight, possibility, clarity, even peace.

The Anatomy of a Prompt

A prompt is not a sentence. It's a structure, built from distinct, modular pieces.

Anatomy of a Prompt

In this book, we'll explore each of these pieces in depth:

1. Request
The request is the ask. The intention. The action you want taken. It lives at the center of the prompt and at the center of the conversation.

2. Examples
Models that shape understanding. Examples show the AI what kind of answer you're after.

3. Format
The shape you want the answer to take: list, email, table, monologue, whatever format the moment calls for.

4. Context

The backstory. The setting. The situation. Without context, even a right-seeming answer can be the wrong move.

5. Role

Who is the model supposed to *be* while answering: A teacher? A CEO? A therapist? A pirate? Give the AI a role.

6. Constraints

Word limits, tone, structure, time period, audience. Constraints don't limit creativity – they focus it.

> > >

You won't need all of them every time, but when you know how to use them – when to add, when to subtract – you can shape conversation with surgical precision or poetic ambiguity, as the moment requires.

Take for example this book, written with the assistance of ChatGPT. The tone is gentle and firm all at once, seven chapters of points that are clear and illustrated with anecdotes. You'll read these stories and get to know the characters here. Through these concepts you may expand your creative avenues and career opportunities. You may even enhance your communication skills. When you're done with the main part of the book you have an appendix of 500 power prompts to offer you a full depth and breadth of potential prompt directions.

This book was conceived and written through a collaboration between ChatGPT and the author, Dan Hermes, through prompts.

Welcome to a world where AIs will teach humans how to interact with AIs.

Do you begin to see how powerful all of this can be?

The final page of this book provides you with the prompt the author used to write it so you may have insight into how this book was created. In the meantime, consider most of what you read here to be living examples of the power of the prompt:

Prompt Power.

And with that, the title of this book, let us begin at the beginning: with the request.

Why We Start with the Request

The center of the prompt is the desire.

Traditionally, the "request" is called the instruction – a term borrowed from programming, technical documentation, and command-line interface culture. But this book is not about issuing orders.

It's about shaping conversations, making meaning in partnership with a powerful mind.

It's about asking questions that matter, in ways that land.

We don't use the word "instruction" in this book, a term from the era of programmatic rules. We use request, because that's what it is:

- A request for clarity
- A request for story
- A request for help
- A request to be heard

And the way you request something shapes what is possible in return.

What a Request Can Be

A request can be bold or quiet. It can be precise or abstract. It can ask for an output... or an insight.

Here are a few types of requests you might make:

Business Requests

"Write a short email to follow up after a sales call."
"Summarize these meeting notes into action items."
"Generate 5 taglines for a startup that builds solar-powered fridges."

Creative Requests

"Turn this short story into a screenplay opening."
"Write a breakup letter in the voice of a 19th-century vampire."
"Help me name a fantasy city built on the back of a whale."

Emotional or Relational Requests

"Help me explain to my partner why I'm feeling distant lately."
"Write a birthday message to my father that feels real, not robotic."
"Give me words to support a friend who just lost their job."

Reflective or Personal Requests

"Help me list the values I actually live by, not just the ones I say I believe in."
"Write a note from my future self, encouraging me to keep going."
"What am I really afraid of in this situation?"

Organizational or Strategic Requests

"Help me turn this chaotic brainstorm into a structured project plan."

> *"Translate this vision statement into something my team will actually care about."*
> *"Find the gap in this business model."*

> \> \> \>

You've seen the breadth of what you can ask (by domain). Now let's look at why we ask at all.

Behind every prompt is something older than code. A reason. A hunger. A spark. Prompts don't begin with the keyboard. They begin with a desire: I want to know... I want to say... I want to try...

Let's walk through the five deeper impulses that drive people to prompt in the first place.

Why We Prompt: Five Human Needs

Behind every prompt is a human reason for asking.
These aren't categories of content. They're modes of intent. Doorways into the act of requesting itself.

These are some reasons we need to prompt:

For Clarity in Professional Life

We seek sharpness. We ask to *see* the idea before it's seen by others.

> *"Generate an opening paragraph for a proposal that conveys urgency, without sounding desperate."*

—the balance of tone starts with the balance of intent.

"Summarize this in three concise bullet points for a team that's been working on it for months (background attached)."

—for when they already know the backstory – give them the path forward.

"Here's the situation: [brief context]. I need a [type of content] that does [goal]."

—use this to stop circling and start building.

For Connection in Personal Moments

We want help finding the words that make meaning between people.

"Help me write a message that asks for [emotional need] without sounding needy or confrontational."

—most emotional clarity is born from tone, not grammar.

"Give me three gentle ways to say: 'I need space.'"

—a direct request doesn't have to be a sharp one.

"Here's what I wish I could say out loud. Turn this into something I could actually send."

—the truth wants a bridge, not a bullhorn.

For Creativity in Expression

We reach for a spark, a strange turn, a new way to say the old thing.

"Act as my creative collaborator. Help me ask for feedback that invites suggestions, not criticism."

—how you ask shapes how others respond.

"Turn this vague idea into a concrete creative brief I can send to a designer."

—dreams need format to survive outside your head.

For Collaboration in Building

We prompt to make together – plans, drafts, ideas, decisions.

"Write a polite but confident request for collaboration that respects the other person's time."

—you're not begging, you're inviting.

For Guidance When We're Lost

We turn to the page with no idea what to say yet. And it listens anyway:

"Help me figure out what I'm really asking for."

—the most powerful requests often begin with a question about the request itself.

> > >

Perhaps you can see why discussion of human needs supersede technical or business discussions in this Prompt 101 primer. The power of

AI is not in its business acumen but in its understanding of the world and the life of humanity within it. Upon that foundation is built disciplines, industries, families, and art. When we prompt we draw upon that vast understanding to help meet our own human needs.

Prompt Power begins with the shape of the ask. The clearer your need, the clearer the door. A request without shape is a whisper into fog. If the request is weak, the format won't save it. The role won't carry it. The tone will be smoke without flame. But if the request is alive – clear, honest, well-shaped – everything else becomes refinement.

The rest of this book will help you to refine. But this chapter, this first turn of the key, is about learning how to ask.

Elena's New Launch

Elena worked in product marketing for a small edtech startup. That morning the team emailed her: "We need skybeam taglines for the new quiz app. Something persuasive. Something that works."

She opened ChatGPT, stared at the blinking cursor, and typed:

> *"Write some persuasive content for our new student quiz app to raise visibility."*

The response wasn't wrong, but it wasn't taglines. It rambled about benefits. Suggested social posts.

She sighed. She hadn't made the wrong request. She'd made a vague one.

So she specified her ask:

> *"Generate five persuasive taglines for an adaptive quiz app for students with learning differences. The tone should be clear, warm, and trustworthy."*

This time:

'Learning, Your Way.'
'Support That Adapts to You.'
'Smart Tools, Gentle Guidance.'
'Study Without Struggle.'
'Confidence Begins Here.'

Better, and now she had something to work with.
It wasn't a smarter model.
It was a smarter request.

What Was Elena Really Asking For?

Elena didn't get a better result because she used fancier language. She got one because, in her second prompt, she asked for something specific. She didn't just say "Write some content," she said:

> *"Generate five taglines..."*

That's a request.

It's the part of the prompt where you say what you want done. Not what it's about. Not what voice to use. Not who the audience is. Those are all important, but they orbit around the request, which is the core action.

This book will show you hundreds of ways to prompt, but they all begin with the verb — the *thing you want done.*

Here are a few:

Create:
   "Write an introduction..."
   "Generate 3 ideas..."
   "Compose a short message..."
   "Design a sample outline..."

Revise:
   "Make this sound more confident..."
   "Shorten this paragraph..."
   "Improve the flow without changing the meaning..."

Translate:
   "Explain this like I'm five."
   "Write this in a less technical manner."
   "Rephrase this with more empathy."

Analyze:
   "Tell me what's missing from this list."
   "What's unclear about this?"
   "How could this land badly?"

Guide:
   "Help me organize this."
   "Give me 3 next steps."
   "How should I start?"

Discover:
   "What am I really asking here?"
   "List possible intentions behind this request."
   "What questions should I be asking instead?"

> > >

These are the actions – the verbs – you're putting into the model's hands. That's what request means. When you prompt, always ask yourself: What do I want it to do? Am I asking it to create, refine, clarify, challenge, or translate?

If that part isn't clear, no amount of tone or detail will save the rest.

Let's see how a request can work in a team.

Elena and the Meeting That Didn't Ask

The marketing war room was full.

Eight team members. Two open laptops. One flickering projector. And zero alignment.

They were prepping for a partner launch – a joint campaign with a big national nonprofit. The stakes were high. So were the expectations. And everyone had ideas:

"We need a statement of values," said one. "An origin story!" exclaimed another. "A headline that makes people care," offered someone else. It was clear they didn't know what they were asking for, and it wasn't because they didn't care. It was because no one had shaped the ask. After twenty minutes of circling, Elena – leaning back in her chair, arms folded – said quietly:

"Can we pause?"

The room stilled. She opened her laptop.

"Let's try something. I'm going to prompt ChatGPT with a real request. Just one. Let's see what we get."

She typed:

"Write a short, mission-driven paragraph introducing a partnership between an edtech startup and a national nonprofit that supports inclusive learning. The tone should be confident but warm. This is for a landing page seen by parents and educators."

A moment later, the model returned:

Together, we're building a future where every student, regardless of ability, has access to learning tools that adapt to their needs. Our partnership unites innovation with inclusion, creating real change where it matters most: in the classroom.

The room was quiet. Then someone said:
"Okay. That's something."

Now they had a starting point. Something to critique. A shape. A direction. A launchpad. The team had been talking around the work. Elena had prompted into it.

Not because she was the best writer in the room, but because she'd made a clear request.

When a Group Talks but No One Asks

Teams often talk in ideas. A request turns talk into action. You don't have to be the loudest in the room to make the smartest ask.

A good request isn't about sounding clever. It's about asking for something real. This is what often happens in teams: everyone brings ideas, intentions, opinions, but no one brings a request. And without a clear request, there is no clear response. What we saw in Elena's meeting wasn't a failure of creativity. It was a failure to form the ask. They needed:

- A paragraph
- For a landing page
- About a partnership
- With a tone that matched the moment

But until Elena named it, no one said that out loud.

Notice that Elena didn't ask the model to "write something great." She asked it to do something small, specific, and actionable. It wasn't about producing a final version, it was about creating a starting point. That's what a good request does:

It gives you something to push off of. It turns discussion into material. It says: "Do this, so we can respond to it."

This Is Why We Start with Request

You could spend hours fine-tuning tone, audience, formatting, but if the action you're requesting isn't clear, you're just spinning words. In the chapters that follow, we'll build on this. We'll explore how examples, format, context, and more shape the conversation. But every strong prompt begins with this question:

What do I want it to do?

That's the heartbeat of the request. And if that's the only thing you master, you'll already be miles ahead of most.

The request isn't just the center of the prompt. It's the gateway to discovering what AI can do for you:

- To reframe a problem
- To diagnose what's missing
- To suggest next steps
- To rewrite, summarize, challenge, restructure, condense, expand, or reposition your work
- To play a role, take a side, show empathy, or give hard feedback

You can ask it to explain something five different ways. Or five times better. Or as if your boss needs it in two minutes and has never read a brief in her life. And if you don't ask, it won't offer.

A vague ask will yield a vague response. Requests are not questions: they are intentions in disguise.

The clearer your need, the clearer the answer.

Seeing the Menu

Before you can become a powerful prompt crafter, you need to expand your idea of what's askable.

You don't need technical skill. You don't need a programming background. You don't need to sound impressive.

You just need to know what you can ask for:

- Action
- Advice
- Strategy
- Persuasion
- Revision
- Reinvention

- Clarity
- Creativity
- Compassion

Let's examine each of these.

Action

Sometimes, you just need it to do something. Not imagine. Not explain. Just do.

> *"Write a subject line for this email."*
> *"Turn this outline into a blog post."*
> *"List five ways to introduce a speaker."*

These are requests that create. They are the digital equivalent of saying, *"Can you do this real quick?"*

When you don't know where to start, ask it to act. Not because it's perfect. But because it's moving.

Advice

You can ask for guidance. Gently. Boldly. Anonymously.

> *"What's the best way to prepare for a raise conversation?"*
> *"How do I announce I'm leaving a job without burning bridges?"*
> *"What are common mistakes when launching a new feature?"*

It doesn't mean you take the answer as gospel, but it gives you something to push against. Something to consider. Advice is a request for wisdom, not facts. It turns the model from a library into a conversation partner.

And sometimes, hearing advice out loud, even if it's not quite right, can help you believe what you already knew.

Strategy

You can ask for thinking. Structured, layered, actionable thinking.

> *"Give me a 3-phase rollout plan for this campaign."*
> *"What's a smart way to increase engagement among lapsed users?"*
> *"Compare two approaches to pricing this product."*

Strategy prompts don't just ask what to do, they ask how to think about it. This is where the model shines: combining known patterns, surfacing assumptions, suggesting frameworks. You don't have to take its plan whole. But you might steal its bones.

Persuasion

You can ask it to help you win hearts, shift minds, or open doors.

> *"Help me write a message that gets someone excited about our product."*
> *"How can I frame this feature as a solution, not a detail?"*
> *"Make this offer feel more urgent without sounding pushy."*
> *"Give me three angles to pitch this to a skeptical CFO."*

Persuasion prompts are about alignment – not trickery. You're not asking it to manipulate. You're asking it to help connect value to need. And sometimes, all it takes is a better angle. A new lead-in. A clearer path between what you offer and what they care about.

That's persuasion. And yes, you can ask for it.

Revision

You can ask it to make what you've written better.

> *"Shorten this without losing meaning."*
> *"Make this sound more confident."*
> *"Fix the grammar and punctuation in this email."*
> *"Change the tone to be more compassionate."*

Revision isn't cheating, it's collaboration. You bring the thought. It helps shape the delivery.

Sometimes you don't need a new idea. You just need the one you had – refined.

Reinvention

You can ask it to change form completely.

> *"Turn this report into a slide deck."*
> *"Rewrite this memo as a press release."*
> *"Take this outline and turn it into a podcast script."*

This is powerful. It's not just wordsmithing – it's format-shifting. A re-invention prompt helps you take what you already have, and see how else it might live. This is especially helpful when you have something half-done but need to adapt it to a new context.

Clarity

You can ask it to make something more accessible and understandable.

> *"Explain this in simple terms."*

"What parts of this message might confuse people?"
"Summarize this in 3 key takeaways."

Clarity is about understanding. And sometimes the most helpful thing you can do is say to the model: "Help me see this better."

Clarity prompts are powerful for revising things you didn't write – a colleague's doc, a legacy policy, a block of messy notes.

You don't need brilliance. You need visibility.

Creativity

You can ask it to play.

"Give me 5 names for a mysterious candle scent."
"Invent a holiday for introverts."
"What would a brand voice sound like if it were a forest?"

Creativity doesn't mean random, it means unusual, expressive, unexpected. You can ask for metaphors. Imagery. Rhymes. Characters. Contrasts. The best creative prompts give a shape and invite surprise.

This isn't about getting the final result. It's about generating raw material you can remix, reject, or rearrange.

Compassion

You can ask it to help you say hard things softly.

"Write a message to a client whose expectations we missed."
"Give me a way to apologize that sounds real."
"How do I respond to a teammate who's grieving?"

This is one of the most underused kinds of prompting, but one of the most powerful. You don't need a robot to feel, but you might need help finding the language for feeling. Compassion prompts are often the most personal, but they also show up at work more than you'd expect. Because so much of our work is with people.

> > >

The model won't always get it right, but it won't get anything unless you ask. And now you know: you can.

Closing: The Core of the Conversation

Every prompt you write begins with a request. Not just because it's the first thing you type – but because it's the center of the intention. What do you want the model to do? That's the heartbeat.

The model doesn't know. It waits. It listens. And the request is what tells it:

"Move. Make. Clarify. Create. Soften. Sell. Summarize. Solve."

This is the most human part of prompting – the desire behind the words.

The rest of this book walks through the tools that make a request work better:

- **Examples** show the model what success looks like.
- **Format** tells it how to shape the response.
- **Context** gives it the backstory it needs to respond intelligently.

- **Role** asks it to step into a persona that fits your situation.
- **Constraints** keep things focused, polished, and useful.

Each one adds clarity to the ask. Each one helps the model not just respond, but respond well. But make no mistake:

Without a clear request, there's nothing for the rest to serve. That is why we start here.

Because everything that comes next – every clever structure, every precise tone, every impressive outcome –

only matters if you've first asked for something worth answering.

The Model Needs a Model

So you've made your request. It's clear. It's honest. It's alive. But the model still has a question:

"What does a good answer look like?"

This is where many prompts fall flat. Not because the ask is weak but because the model has nothing to model.

This is where you stop telling and start showing.

In the next chapter we'll explore how to provide examples so the model isn't just guessing at what you want, but drawing from patterns you've given it.

Because some of the best prompts don't just ask for something.

They show what success looks like.

CHAPTER 2: GIVE EXAMPLES

If the request is the command, the example is the lesson. Nothing guides an AI more powerfully than a demonstration. This chapter shows you how to build your prompts around showing, not just telling.

We call it Machine Learning (ML) because AIs identify and learn from patterns, which we call training. That makes prompts work particularly well when they include examples.

Examples don't constrain the model – they give it guideposts. They tune the output to match your voice and your rhythm. They say: "Do it like this." And the model listens.

How the Model Learns From What You Show

You've asked for something.
The model is listening.
But it still doesn't know what you mean.

You said "make it friendly."
You said "a sharp summary."
You said "help me pitch this."

But what does "friendly" sound like?
What does a "sharp summary" include?
What kind of pitch do *you* think works?

This is where many prompts fail – not in the asking,
but in assuming that the model already knows what success looks like.

And that's where **examples** come in.

Tell Me vs. Like This

An example says:

I don't just want a good answer.
I want a good answer that looks like ***this***.

It might be:

- A sentence you've already written
- A fake example with the structure you want
- A tone sample, a snippet, a prototype
- Even your improved rewrite of what the model just gave you

You're showing it a pattern.
And language models thrive on pattern.

You don't have to know exactly what you want.
But if you can show something close – something half-right,
something shaped – you're giving the model a mirror to reflect from.

A sample. A frame. A hint.
That's all it needs to begin improving.

The Apology

It was a small mistake – a broken link in a campaign email. But the customer it affected was a large one, and their reply came in hot: "This isn't the first time. We expect better."

The team gathered in a virtual huddle. Heads down. Tones clipped. Elena was there again – tasked with drafting the reply.

They started with the basics:

"Say we're sorry."
"Make it professional."
"Keep it short."
"Don't over-apologize. But don't underdo it either."

Elena typed a prompt into ChatGPT:

> *"Write a short professional email apologizing for a broken link in a customer campaign. Be warm but not overly apologetic. Keep it respectful and concise."*

The response came back fast, and it was... fine:

> We regret the inconvenience caused by the broken link in our recent campaign. We've addressed the issue and are taking steps to prevent it from happening again. Thank you for your understanding.

Nobody hated it, but nobody wanted to send it either. "Still feels stiff," someone said. "Yeah, it doesn't sound like us," said another. Elena sighed. She had asked for the right thing, but she hadn't shown it yet.

So she copied an email they'd written months ago, sent to a partner when something similar had gone sideways. It had the tone they wanted: Direct. Thoughtful. Not too polished. Just real. She added it to the prompt:

"Here's an example of our voice in a similar situation:

'Thanks for flagging that. We definitely missed the mark, and we appreciate your patience. We're correcting it now and we'll be following up with an updated version shortly. Let us know if you spot anything else – we're watching it closely too.'

Use this as a tone reference and rewrite the apology email to the customer who received the broken link."

The model tried again:

We really appreciate you pointing this out. The link issue slipped through and we've already fixed it. We're keeping a close eye to make sure it doesn't happen again. Thanks for your patience – and for holding us to a higher standard.

This time, the reply felt like it came from *them*. The team nodded. "Yep. That's it."

Send.

Takeaways

- Even a perfect request may fail if the model doesn't know what "good" looks like.
- You don't have to provide a perfect sample – just a clear one.
- Examples aren't cheating, they're collaborating.

Elena didn't change the request. She didn't rephrase it. She didn't clarify the audience.

She showed an example.

The model didn't have to guess what she meant by "professional" or "warm." It saw it. It heard the voice she wanted. It mapped her tone, her phrasing, her rhythm. Language models can mimic almost any writing style – if you give them something to mimic.

That's what an example does: It creates a pattern.

You can drop in a sample message. A sample phrase. Even just a list of adjectives you like. And the model will say:

"Ah. You want something like this. Got it."

Example Types

Let's walk through a few common types of examples you can include in your prompt – and what they're best for.

1. Tone Example

A snippet that captures the voice you want

Use this when you want the model to match a vibe – playful, confident, bold, academic, whatever.

> *"Here's the kind of tone we usually use:*
>
> *We know security's complicated. That's why we make it simple. We handle the hard parts so you don't have to.*

> *Use this tone to write a landing page for our new compliance product."*

Tone examples are often short, just 1–3 sentences, but powerful. They tell the model how to sound.

2. Structural Example

A sample that shows the format or flow you want

Use this when you want the model to follow a pattern – like a list, a framework, or a style of explanation.

> *"Write a thought leadership post in the same structure as this one:*
>
> *- Start with a question.*
> *- Share a surprising insight.*
> *- Use a metaphor.*
> *- End with an invitation.*
>
> *Use this structure to write a post about burnout in startup founders."*

Structure examples give the model a skeleton to build upon. They're especially useful for content creation.

3. Before and After

Show what you don't want – and then what you do

This is especially helpful for revising or rewriting prompts.

> *"Here's the version we don't want:*
> *We are reaching out to you in regards to the following issue...*

And here's what we prefer:
Just wanted to flag something quickly – we noticed a small issue
and we're on it.

Rewrite this message to match the second style."

This helps the model *course correct* toward your ideal.

4. Self-Correction Example

Take what the model gave you – and rewrite it better

Sometimes, the best example comes from your own edits.

"You wrote this:
We understand your frustration and are working to address the
issue.

I rewrote it like this:
Thanks for calling this out. We're on it – and we'll follow up
shortly with an update.

Use my version as a tone reference for future responses."

This teaches the model by showing how your version differs from its
guess.

5. Imaginary Sample

Make up an example to guide the model

You don't always need a real-world quote. You can write a made-up
one, just to create a pattern.

"Write a customer testimonial that sounds like this:
'The interface is gorgeous, but what really sold me was the support
team. They're like having tech whisperers on call.'

Generate 5 more in this tone."

6. Instructional Examples

A sample paired with an explanation of why it works

This one is great for onboarding your team's voice or rules.

"When we respond to customer complaints, we use this format:
1. Acknowledge the issue
2. Take responsibility if it's ours
3. Outline what we're doing
4. Thank them for their patience

For example:
'You're right – we missed that. It's been fixed, and we've updated
the docs to make sure it doesn't happen again. Appreciate you
flagging it.'

Use this format to respond to a user who found a bug in the
mobile app."

You're not just giving an example – you're teaching with it.

7. Bad Examples

Provide a bad example and ask the model to avoid it

This is especially helpful when tone is critical.

"Don't write like this:

We apologize for the inconvenience and regret any negative impact this may have caused.

It's too cold and robotic. Make it sound more human."

Bad examples set a boundary: "Anywhere but there."

8. Persona Example

Examples generated by a specific person or group

"Our CEO tends to write like this:

'We're learning fast, and we'll get it right. That's a promise, not a platitude.'

Use this voice to respond to a critical investor email."

These examples help the model match point of view.

> > >

Those are some of the most common example prompts: Tone, Structural, Before and After, Self-correction, Imaginary, Instructional, Bad, and Persona. More examples can be better than fewer but the important thing is quality.

You Just Need One That Lands

You don't need to give five examples. You don't need a whole doc. Because once the model sees how you want it to sound, it doesn't have to guess. Sometimes one Is enough, but five paints a picture. When you include one example in a prompt, you give the model a signal. When you include several, you give it a pattern. That's the difference between guessing the vibe – and locking it in.

Multiple examples help the model:

- Spot consistency
- Weigh what matters most
- Avoid outliers
- And generate results that feel intentional, not random

Multiple examples like this:

"Here are five taglines from past campaigns we liked:

- Clear Power for Complex Systems
- Control Without the Chaos
- Every Detail, Under Control
- Power, Made Predictable
- Nothing Slips Through

Generate five new taglines for our next product launch. Keep the tone similar – crisp, direct, quietly confident."

That's not just prompting. That's training. You're not hoping it understands your taste – you're *showing* it.

Multiple examples like this: (get it?)

"Our support team tends to sound like this:

> - *'Totally fair point. Let me dig into that and I'll follow up by EOD.'*
> - *'Appreciate your patience – this one's on us. Fix incoming.'*
> - *'Good catch! Flagging that for the product team now.'*
>
> *Using this tone, respond to a user who reported a broken button."*

This doesn't just produce a usable response. It helps your AI co-worker speak in your voice.

If you give five examples that sound wildly different, the model won't know which way to steer. But if they share tone, structure, or rhythm, it can match the pattern beautifully. More cohesive examples = more certainty. Especially when you're going for tone matching or voice generation at scale.

Here's Elena mentoring a co-worker on how to do it.

Elena and the Junior Marketer

The meeting had ended. The email was sent. The tension in the room – dissolved. But Elena noticed Jake lingering.

He was new. Smart. Eager. And clearly frustrated. "You made that look easy," he said, "But I asked ChatGPT for the exact same thing earlier and it gave me junk."

She smiled, "Did you show it anything?"

"Well… no. I told it what I wanted."

"Right," she said. "But sometimes you need to show, not just tell."

They sat down in a side room, two laptops open, coffee between them. "What were you trying to write?" she asked.

"A follow-up message to partners who came to our demo. I wanted it to be short and appreciative, but with a clear next step."

"Got the prompt?" she asked. He pulled it up:

> *"Write a short follow-up email thanking partners for attending our product demo and encouraging them to sign up for a trial."*

It wasn't a bad prompt. The AI's reply was something like:

> Thank you for attending our recent demo. We hope you found it informative. If you haven't already, we encourage you to sign up for a trial today.

It was the kind of email you delete before the second sentence. "Now watch this," Elena said. She dug into a folder and copied in three messages they'd sent after past events. They had voice. Energy. Personality.

She pasted them under the prompt:

> *"Here are three examples of the tone we usually use:*
>
> *'Thanks again for coming to the demo — great questions, by the way. If you're curious what it's like to actually use the tool, hit the link below and try it free for 14 days.'*
>
> *'Loved having you join us. This tech only makes sense when you get your hands on it — grab a free trial and see what we mean.'*
>
> *'Appreciate you showing up! If you're ready to test-drive what we showed off, here's your chance — free for two weeks, no strings.'*

Write a new follow-up email using this tone and structure."

The model responded with something Jake hadn't expected:

Thanks again for joining our demo. Want to try it for yourself? You can start a free trial today – no setup, no commitment. We'd love to hear what you think.

Jake stared, "That's actually... kind of great."
"Right?" Elena said, "Because we showed it what great sounds like."

Takeaways

- Examples don't just improve the output – they teach the model how you think.
- If your prompt is falling flat, don't just say more. Show more.
- One good example is a flashlight on the subject. Three is a floodlight.

Jake had a decent request. The model responded with a passable email. It sounded like... everyone else. Because one of the most underappreciated truths in prompt crafting is this:

The model doesn't default to your voice. It defaults to the most common voice.

And most common isn't what your brand wants. It's not what your company rewards. It's not what makes anyone click.

That's why multiple examples are so powerful. Especially when:

- You're writing in a distinctive voice
- You're producing repeatable content at scale

- You want tone consistency across a team
- You're handing off drafts to others (or to the model itself)
- You want the model to learn, not just respond

This is how voice gets trained — Not with rules, but with recognizable repetition.

Elena didn't just prompt better. She showed Jake that a good prompt isn't only about asking, it's about modeling, and that's what examples are at their best: Small models of what good looks like.

Just as Elena provided a great example for Jake, a coach must do the same for their players.

A Tale of Two Coaches

Both teams ran the same drill that day.
    Same field.
    Same sweat.
    Same burning sun.

On one sideline, a coach barked:

"Run it again!"

And the player did.
Fast. Wild. Heart in it, but guessing.

"No!" the coach snapped. "That's not what I meant!"

"Then what *did* you mean?"

The coach waved his hand and said it louder. Said it harder.

"Run it again! This time get it right!!"

So the player kept running.
Kept missing.
And the drill dissolved.

On another sideline, the coach of the winning team watched.
When her player asked:

"How do you want me to run it?"

She didn't raise her voice.
She raised her hand and traced the shape of the movement in the air.

The player smiled, nodded once, and moved like he'd already seen the success in a dream.

And the drill —
the same one as on the other field —
became poetry in cleats.

The scoreboard wouldn't show it yet.
But that moment was the win.

Post-game Analysis

A coach yelling "do it better"
is no different than a boss saying "write me something great,"
or a prompt that says "Do this."
Without example, there is no aim.
Just volume.

Without examples, there's no shared reality.

We speak the same language,
but imagine different outcomes.
And that's where things fall apart.

The coach who wins isn't the one who yells louder.
It's the one who gives their desire a shape.

Reflection

You can write the best request in the world –
clear, honest, focused –
and still get something that feels *off*.

Because the model doesn't just need to know what you want.
It needs to know *what good looks like.*

That's what examples are for.

They're not instructions.
They're impressions.
Mirrors that let the model see what you see.

You've seen the power of:

- **Tone Examples** – A few lines of voice can shape a whole paragraph
- **Structural Examples** – Give it a frame, and it fills it well
- **Before and After** – Teach by contrast
- **Self-Correction** – Edit what it gave you, then say: "Do it like this."
- **Imaginary Examples** – You don't need real data to make a useful sample
- **Instructional Examples** – Spell out the format and show it.

- **Bad Examples** – What not to do.
- **Persona Examples** – Show it how *that person* might say it

You don't always need five.
Sometimes one good example is enough.
But where tone and trust matter –
more quality examples mean less risk.

Because the model doesn't guess style.
It copies it.
And it can only copy what you give it.

Shape the Shape

You've made your request.
You've shown what good sounds like.
But the model still wants to know:

"How long should this be?"
"Should it be a list? A letter? A chart?"
"Where should I stop?"

That's where format comes in.
It doesn't change *what* you're asking for.
It changes how the answer appears.

And that can make all the difference.

Ready to shape the shape?

CHAPTER 3: HOW TO FORMAT YOUR OUTPUT

It's not just about content – it's about *shape*. Format is the shape of a prompt's output.

This chapter explores the power of formatting prompts to direct layout, sequence, rhythm, and even emotional resonance. From slides to scripts, tables to outlines – these are the forms that comprise information.

Let us turn now to a format of yore in a tale inspired by Jorge Luis Borges.

The King and the Mapmaker

The king summoned the mapmaker to the great hall. He was tired of being surprised by border raids and unruly rivers.

He pointed to a blank table and said: "Make me a map of the kingdom."

The mapmaker bowed: "What kind of map, Your Majesty?"

"The whole thing," the king said, waving. "All of it."

So the mapmaker set upon the charge with obsessive devotion. He measured every stream, every barn, every alley. He drew every stone wall, every goat path, every elevation. In its sublime fastidiousness, the map, when finished, was surprisingly large.

It was, in fact, the size of the kingdom itself.

The king stared at it, gobsmacked: "Mapmaker, what have you done? I can't use this. I can't carry this. No one can! What good is a map you can't hold in your hands?"

The mapmaker blinked, and reminded the king of his request:

"You said the whole kingdom."

The mapmaker lived out his days in the dungeon.
The king remained lost in his own kingdom.

Here Be Dragons

The King had content — too much of it. He had land, lords, livery… even loyalty.

But what he lacked was form.

He didn't need more information. He needed information he could use.

That's where the mapmaker came in: He didn't invent the world. He formatted it.

The king needed for him to take something vast and make it *holdable*.

That's what format does in a prompt.

Lest we meet the mapmaker's fate, we must ask ourselves:

"What am I really asking for – and what should it look like when it arrives?"

When you tell the model what shape you want your map to take, you do more than make it look nice.

You:

- Make it easier to use
- Facilitate decision-making
- Make it more satisfying to read
- Make it easier to share

Because how something is presented matters as much as what's inside: A list. A summary. A grid. A table. A single sentence.

The right format doesn't just display your request. It translates it into impact.

Into a map a king or queen could use to defend and develop the kingdom.

Format a Useful Map

If the king had a keyboard instead of a crown, his prompt might have looked something like this:

> *"Summarize the territory I control, including trade routes,
> geography, rival region borders, and alliances, into a one-page
> visual map I can study quickly."*

That's a format-driven prompt.

He isn't asking for more knowledge. He's asking for presentation.
Sometimes what you want is already known –
but it doesn't feel known because it's not in a usable shape.

The model can produce almost anything.
But if you don't tell it what form you want,
you're asking it to pour water without a glass.

Format is the glass.
The bowl.
The blueprint.
The slide deck.
The tweet.
The list of ten.
The table with three columns and ordered rows.

It's the delivery mechanism for the idea. And when you ask for it well,
you don't just get content. You get clarity.

When we write: *"Tell me what to do."* The AI might give you:

- A bulleted list
- A philosophical treatise
- A sarcastic haiku

—when all you want is a checklist. Or a short email. Or a pep talk. But
you don't say the format.

The machine – like the player – just *runs it again*.

Here are some types of format prompts, for when form matters as much as meaning.

Format Types

Prompts can mold content into many shapes using formats. Here are some examples:

1. List Format

Prompt:
"List 5 creative ways to open a leadership retreat."

Output:
1. Start with having a blindfolded team walk into the room, then reveal a shared challenge.
2. Use an anonymous pre-written "fears and hopes" box and read a few aloud.
3. Begin with the most difficult question the group will face – no small talk.
4. Share one-minute audio messages from team members' colleagues saying what they admire about them.
5. Start in silence: give everyone one sticky note, one word, and one minute.

2. Paragraph Format

Prompt:
"Write a short paragraph explaining why leaders should embrace vulnerability."

Output:

Leaders who show vulnerability invite trust. By acknowledging uncertainty or failure, they create psychological safety – and signal that it's okay to be real, to be human. This doesn't weaken authority – it deepens connection and resilience.

3. Table Format

Prompt:

"Compare the risks and benefits of in-person vs. remote employee onboarding in a table."

Output:

Format	Benefits	Risks
In-Person	Builds rapport quickly, stronger culture	Travel costs, limited flexibility
Remote	Scalable, more accessible	Harder to build trust, distractions

4. Headline / One-Liner Format

Prompt:

"Give me 3 headline options for a blog post about AI in education."

Output:

1. Smarter Classrooms Start With Smarter Questions
2. AI Won't Replace Teachers, But It Will Change Them
3. Rewriting the Curriculum – One Algorithm at a Time

5. Multi-Part Structured Format

Prompt:

"Generate 2 taglines for a mindfulness app. For each, include the tagline, a rationale, and the intended audience."

Output:

1. **Tagline:** Quiet the Noise, Find Your Rhythm
   Rationale: Emphasizes calming internal chatter while suggesting personal pace.
   Audience: Burnt-out professionals, ages 30–45.
2. **Tagline:** Stillness Is a Superpower
   Rationale: Turns rest into strength, using superhero metaphor.
   Audience: Ambitious achievers new to mindfulness.

6. Slide / Outline Format

Prompt:

"Turn this paragraph into a 3-slide outline:

'Our company is focused on sustainable logistics, partnering with urban delivery services to reduce emissions. We've deployed AI routing, reduced idling times, and incentivized low-emission fleets.'"

Output:

Slide 1:

Our Sustainability Mission
- Urban delivery partnerships
- Emission reduction focus

Slide 2:

Tech That Powers the Shift
- AI routing optimization
- Idling time reduction

Slide 3:

Incentives for Greener Fleets
- Tiered rewards system
- Long-term impact forecast

7. Creative Format

Prompt:

"Explain blockchain as a bedtime story for a curious 7-year-old."

Output:

Once upon a time, there was a magical notebook that everyone in the kingdom could see, but no one could erase. Every time someone bought a toy, shared a snack, or traded marbles, the notebook wrote it down – so everyone knew it was fair. That notebook was called the *blockchain*, and it helped the kingdom stay honest, even when the lights went out.

> > >

Each of these format types shapes how the model delivers the answer – not just in style, but in usefulness. You're not just asking it to talk. You're asking it to organize, present, and persuade.
You're asking for it to become easier to understand, use, share, or sell.

The format you choose depends upon:

- Who will read it
- How much detail they need
- What they will do with it

And the clearer you are about that, the sharper the model's answer will be.

But just as format fell short in the king's castle, format can fumble in the meeting room.

The Invisible Insight

It was his first strategy presentation. Nico had done the work – deep work. He pulled numbers from three systems, tracked customer churn across six months, talked to sales reps, and distilled it all into one sharp insight:

"Our onboarding sequence was bleeding users before day three."

This was real. Important. Fixable. He set out to make his case and ring the alarm bells. His ChatGPT prompt read:

"Write a summary of our user onboarding process, drop-off rates, and possible improvements based on internal interviews and Salesforce data."

He attached his Salesforce documents and the model delivered – a tidy, professional, five-paragraph block of strategy analysis. It was good.

Too good. No bullets. No highlights. No callouts. Just text. Which he copied straight into slide 4.

At the meeting he spoke over it – adding more explanation.

The execs squinted. Someone flipped backward a slide.
One nodded like a human screensaver. And then... they moved on.
No one pushed back. No one asked for follow-up. No one remembered the insight.

Nico had found the leak. Named it. Outlined the fix.
But he never asked the model:
"Help me shape this in a way they'll actually see."

No formatting. No contrast. No takeaway. No friction. No fuel.
Just... copy.

And the worst part?
Nico walked out thinking that maybe it hadn't been worth presenting at all.

Takeaways

Nico wasn't wrong.
He found a meaningful pattern in the data. He identified a clear cause.
He even prompted the model clearly and concisely.

The request was solid. The insight was real.
But insight doesn't land if no one can *hold* it, like the map.

That's the brutal truth of communication:
If you don't shape your answer, it could become invisible.

Nico forgot that execs don't read paragraphs.
They scan. They prioritize. Context over content.

He needed to say: "Help me format this so they see it fast."
Or: "Turn this into a 3-slide outline with bold headlines and clear
next steps."
Or even: "Summarize this insight in one sentence, one number, and
one call to action."

With a few extra words, Nico could've asked for:

- A key stat in bold at the top
- A 3-part slide structure: Problem, Insight, Fix
- A short list of recommendations
- A one-sentence summary that sticks in the room

Even better: "Write a slide headline that would make someone want
to pause and ask a question."

That's not formatting for style.
That's engagement.
It's not about dumbing it down.

It's about framing it up.
Because smart people still need help seeing what matters.
And the better your format, the better your signal gets through the
noise.

After Nico's meeting, Elena did the sales strategy presentations.
Nobody said it out loud. But when the next deck needed a clear story,
clear structure, clear slides…
It was her name they put on the calendar.

And Nico? Well – he got better at making himself heard. Eventually.

Reflection

Clarity isn't just about content.
It's about what the eye sees first.
What the mind grabs hold of.
What the room *remembers*.

We've seen how format can:

- Make ideas **scannable**
- Turn paragraphs into **slides**
- Distill complexity into **a table**
- Translate insight into **a one-liner**
- Bring a voice to life as **a fable, a tweet, a script**

We've also seen what happens when it's missing –
when good thinking gets buried
because it was presented like a brick instead of a door.

The prompt isn't just about the what.
It's about making the what *usable*.

Backlit

You've asked for something.
You've shown what it should look like.
You've told it how to shape the answer.

But there's still a question lingering in the model's mind:

"What am I supposed to know before I begin?"

That question leads us to context –
the background, the backstory, the ingredients behind the request.
Because a model can be clear and clever...
but if it doesn't have context,
it's guessing in the dark.

Let's turn on the light.

CHAPTER 4: WHAT'S THE CONTEXT?

Context is the invisible lens. It tells the model not just what to say, but why and for whom. Without context, prompts are blunt. With it, they become surgical. You don't always need long instructions – just smart ones that let the model "see" what you're doing. For example, here are some statements:

"That's the third one this week."

Is that good? Is it bad? Is it too many?
Are we talking about sales… or broken chairs?

"Okay… let's send it."

A green light? A gamble? A joke?
Did someone finally finish something, or are they giving up?
It could mean *hope it doesn't crash.*

"She said yes."

To a raise? A job offer? A second date?
Is this the start of something new or years in the making?

DO THOSE STATEMENTS MAKE SENSE?

Of course they don't.

Because without context, language becomes a screen with the brightness turned down.

The words are there. But the meaning hasn't loaded.

Let's turn the lights back on:

"That's the third one this week."
Context: Three clients signed contracts after the new demo went live. That's three new accounts, three wins, and the best week the team's had in a quarter.

"Okay… let's send it."
Context: After four weeks of building, rewriting, and debating every headline, the growth team is finally launching the rebrand. And yeah, they're nervous. But they're ready.

"She said yes."
Context: After dinner, on the walk home, under the quiet hum of a streetlight. They'd been together five years. He asked. She smiled. And said yes.

> > >

Context is the frame.
It's not the picture, but it helps you see it clearly.

Without it, prompts fall flat.
With it, even short requests can hit the mark.

What Is Context in a Prompt?

Context is background information that helps the model interpret your request. It's everything the model needs to know before it starts answering.

It answers questions like:

- What's this about?
- Who is this for?
- What's already happened?
- What should be included, avoided, or emphasized?
- What matters here?

You can think of context as the quiet paragraph before the ask.
It says: "Just so you know…"

Context Can Be:

- A business situation
- A snippet of dialogue
- A product description
- A summary of a meeting
- A link to prior performance
- A note about the audience
- A document you uploaded
- A user persona
- A vibe

Context can be one sentence.
Or it can be an attached strategy deck.

The more relevant and focused the context,
the **smarter** the response.

Elena in the Briefing Room

Elena's ingenuity landed her the top communications role at a rising energy startup. They were smart. Funded. Scaling faster than their strategy. And this morning, her CEO had cornered her outside the elevator with a familiar bombshell:

"Can you get me a killer narrative for the investor deck? Something fresh. We need to punch above our weight."

No timeline.
No brief.
Just *punch above our weight.*

Elena nodded, walked to her desk, and opened a blank doc.

Ten minutes later, she opened ChatGPT and typed:

"Write a compelling narrative for a cleantech investor pitch."

The result was… fine.
A couple of buzzwords. Something about disruption.
She could almost hear a roomful of distracted VCs checking their watches.

She deleted it.
Took a breath.
Then did what she'd been learning to do.

She prompted again:

> *"Here's the situation: We're a Series A cleantech company with a working prototype but no revenue. We just got picked up by a national accelerator. We're asking for $12M to expand operations. The CEO is visionary, but we lack traction. We need to signal future dominance without overselling.*
>
> *Write an investor narrative that frames us as a high-risk, high-insight bet — 'We saw it before others did.' Make it tight and charismatic."*

This time, the response gave her something she could feel.
It told a story about identifying shifts in grid infrastructure before the big players caught on.
It made the risk sound like insight.
It gave her the spine of a speech she could reshape with the CEO's voice.

Later that day, she used a variant of the same prompt to generate:

- A 3-slide summary
- A 20-second verbal pitch
- And a LinkedIn teaser for the launch announcement

What she'd really written wasn't copy.
It was clarity.
She hadn't just asked for *words* — she'd asked for *what the words needed to do.*

And that made all the difference.

Takeaways

- A vague ask yields a vague answer – even if the words sound right.
- The more context you bring to your request, the less rework you'll face.
- A good request tells the model **why the words matter** – not just what kind you want.

A request is not a suggestion. It is an invocation.

The moment you type your prompt, you are not just telling ChatGPT what to do – *you are shaping how it thinks.*

How to Include Context in a Prompt

There are three primary ways to include your context into a GPT prompt. We're not talking about a sophisticated automation or enterprise repository setup where all kinds of timely and appropriate data could be automatically and invisibly included in a prompt, often called Retrieval Augmented Generation (RAG). We're talking about your powerful fingers typing characters and words directly into the GPT prompt on your screen.

1. Inline Context

Put the background directly into the prompt:

> *"We're launching a budgeting app for college students. It rewards small wins and celebrates consistency, not just big goals.*
>
> *Write a landing page headline that captures that tone."*

The model knows the request.
But now it knows the product, the values, and the vibe.

Note that in this book, we **never** assume the model has context unless
we show you how it's clearly provided.

2. Uploaded Context

If you're using ChatGPT with file support, you can say:

> *"I've uploaded our Q3 customer feedback report. Based on that,
> suggest three new onboarding improvements for the mobile app."*

The context lives outside the prompt, but informs it directly (with
uploaded documents, tables, and PDFs).

If you don't upload it or paste it, it doesn't exist.

3. Running Context

Sometimes, you can build context over time:

You: "Here's what we offer..."
You: "Here's the audience we're targeting..."
You: "Okay, now write a short ad using that info."

This kind of "threaded context" works *in-session.*
But the moment you refresh, sometimes it's gone.
So in best practice?

Include what matters every time.

The Rule of Context

If you want the model to respond like it knows what you're talking about, you have to tell it what you're talking about.

Assume nothing.
Supply everything that matters.
Keep it tight. Keep it relevant. Keep it close to the request.

Because the model isn't psychic.
But it *is* brilliant – when it knows where to start.

The Letter She Couldn't Start

She had been staring at the blinking cursor for hours.
The screen was blank. The memories weren't.

Elena hadn't spoken to her sister in over two years – not since the funeral, not since the argument about the house.
Now, she wanted to say something.
Not everything. Not an apology, not exactly.
Just something that could begin a thaw.

She opened ChatGPT and typed:

> *"Write a message to my sister."*

The model responded instantly:

> Hi there! Hope you're doing well. Just wanted to check in and see how things are going!

Elena stared at the chipper nothing on her screen.

She deleted it.

She tried again:

> *"Write a message to reconnect with my sister after a long silence."*

This time it came back soft. Too soft.

> It's been a long time. I hope you're doing okay. I'd love to
> catch up sometime.

It wasn't wrong. But it didn't sound like her.
It didn't carry the ache, or the hesitation, or the small shard of pride
she hadn't yet dislodged.

She sat back, frustrated.
And then – without fully realizing it – she began to *prompt differently.*

> *"Here's the situation: I haven't spoken to my sister in two years
> after a family falling out. I want to send a message that feels
> honest but gentle, with just enough warmth to invite her in. I
> don't want to get into everything – we're not ready. I just want to
> open the door.*
>
> *Write the message in my voice. I tend to be concise, a little dry, but
> sincere."*

The response came back slowly.
Measured. Cautious. Familiar.

> Hey. I've been thinking about you. I know there's a lot we
> haven't said – and maybe still shouldn't – but I'd really like
> to reconnect, even just a little. No pressure. Just letting you
> know the door's open.

Elena read it once.
Then again.
Then added one line of her own:

Only if you want to.

She hit send.

And for the first time in a long while,
it felt like a conversation had begun.

Why the Third Prompt Worked

Elena's third prompt succeeded because it named the relationship.
It gave emotionally anchored details and defined the desired tone,
asking to "open the door" without "getting into everything".

Elena's sharing gave the model context: "Here's the situation..."

The model responded with the words she needed to move forward
with her sister. Move towards the start of something true.

Chapter Reflection: Context

You've learned how to ask. You've learned how to show. You've learned
how to shape.

But none of it works if the model doesn't know what's behind the
prompt.

Context is what brings a prompt into the real world. It turns ideas into
understanding. It fills in what the model can't guess.

Who's it for? What's at stake? What's the moment? What matters?

Context answers these questions so the model can respond like it knows where it is.

We've seen it at work:

- In launch campaigns
- In a sensitive email
- In requesting guidance
- In pitches, slide decks, and summaries

Context doesn't need to be long but it does need to be present.

It's the map before the mission.

And Now the Model Asks: "Who Am I?"

You've told it what to do. You've shown what it should look like.
You've shaped the form. You've shared the world.
But there's one more question the model is ready to ask:

"Who should I be?"

That's where we go next.
The chapter on role – where voice, tone, and perspective come to life.
Because sometimes, it's not what the model says.
It's who's speaking.

CHAPTER 5: YOUR ASSISTANT'S ROLE

Telling the model what it is, who it is, or how it should behave unlocks a different class of interaction. This isn't just tone – it's perspective. By assigning a role, you invite the model into character.

You can prompt it as a chef, a coach, a philosopher, or a project manager. Each identity reshapes the answers it gives – even when the topic stays the same.

> *"Pretend you're my therapist."*
> *"Act like a skeptical investor."*
> *"Speak as if you were Steve Jobs."*
> *"Write this like a friend who's disappointed but still hopeful."*

These aren't gimmicks.
    They're invitations.

Invitations for the model to take on a role –
to shift its posture, its tone, its judgment, even its values.

And when used well, they can turn a flat answer
into something alive.

The Model Doesn't Have a Voice – Until You Give It One

GPT doesn't speak like a person. It speaks like a million people. Which means its voice is neutral, until you anchor it.

That's what role does. It's how you ask it to:

- Speak with authority
- Offer care
- Use skepticism
- Stay playful
- Get formal
- Get human

Not just what to say.
How to say it.
Who to say it as.

Why Role Matters

Every real-world conversation happens in context and in character. You talk differently when you're:

- A manager giving feedback
- A best friend calming nerves
- A product lead in a roadmap meeting
- A brand, a voice, a mood

Without role, the model may sound too stiff. Or too casual. Too salesy. Or not persuasive enough. Too confident. Or not confident *enough*.

But when you give it the right role, you get responses that don't just make sense – they feel right.

The Pirate and the Sandwich

They were supposed to be doing homework.

But the math sheet lay abandoned on the floor, and ChatGPT was open on the laptop. Ben, age 10, sat at the keyboard. Caleb, age 8, was draped across the bed, holding a peanut butter cracker like a microphone.

"Make it talk like a pirate," Caleb said. Ben laughed like he'd been waiting all day for that request.

He typed:

"Tell us how to make a sandwich, but do it like a pirate."

ChatGPT answered:

Arrr, first ye be takin' two slabs o' bread – none o' that soggy bilge, mind ye – and layin' 'em down like the planks they are!

Caleb shrieked. Ben kept typing:

"Now do it like a robot."
"Now like a poet."
"Now like my big sister when she's mad."

Each voice brought new words. New rhythms. New choices.
Same sandwich.

At some point, they forgot to laugh. They were reading silently. Listening. Later that night, when their mom came in, she asked if they'd finished their math.

"Almost," Ben said, "but... did you know sandwiches can sound like anything?"

What the Boys Were Really Learning

Ben and Caleb weren't playing with lunch instructions. They were playing with voice.

They were discovering – through laughter and language – that *how* something is said changes what it feels like.
Even when the facts stay the same.

Each time they changed the role – they shifted the tone, the pacing, the intention.

The pirate gave commands. The robot was sterile. The poet felt it. His big sister meant it.

Same request. Different voice. Different effect. They were learning the lesson professionals often forget:

Voice shapes meaning. Role shapes response.

> > >

Now, let's turn that play into power. Now that we've seen what two kids can teach us about role, let's step into the real world again, where voice isn't just entertainment – it's strategy.

Here's a spread of practical, real-world role prompts, each one showing how the same request changes when the model is asked to speak differently:

1. The Coach

Prompt:
"Give me advice on how to handle a rough week at work. Speak like a coach who believes in me but won't let me off the hook."

Response:
You had a rough week. Fine. You're allowed one. But now we get up. What did you learn? What will you do differently Monday morning? Take thirty minutes, make a plan, and then get out of your own way.

2. The CEO

Prompt:
"Explain this new product to potential investors. Speak like a confident, visionary CEO."

Response:
We're not just building a tool – we're redefining how companies think about knowledge flow. What we've created is a way to turn scattered expertise into real-time intelligence. And the market is begging for it.

3. The Skeptic

Prompt:

"Evaluate our new marketing tagline. Act like a skeptical customer who's seen it all."

Response:

'Smarter solutions for a better tomorrow'? Seriously? Sounds like ten other SaaS companies I ignored this week. What makes yours different – and why should I trust you?

4. The Therapist

Prompt:

"Help me reframe my anxiety about switching jobs. Use the voice of a compassionate but honest therapist."

Response:

It's completely normal to feel fear in the face of change. But fear doesn't mean you're wrong – it means something important is happening. Let's unpack where the fear's coming from, and what it might be trying to protect.

> > >

Each of these prompts asks for roughly the same thing: help, advice, explanation, feedback.

But the voice changes the *relationship*. The model stops being a neutral observer. It becomes a coach, a confidant, a leader, a skeptic, a teacher, a narrator.

That's the magic of role. That's why role matters.

How to Assign Role in a Prompt

Assigning a role means explicitly telling the model who it should pretend to be while answering.

The model will answer any question you give it. But unless you tell it who to be, it won't know how to speak. Role is the part you cast the model in.
It's the voice it wears, the posture it takes, the kind of mind it pretends to have while it speaks.

It's easy to assign. You simply say:

> *"Speak like a mentor."*
> *"Answer like a friend."*
> *"Respond as a therapist, kind but direct."*
> *"Explain this like a tired professor who still cares."*

You can place that request anywhere in the prompt. At the beginning, in the middle, or right at the end. What matters is that you say it – clearly. Because the model won't remember your tone for long. It might carry over a turn or two, but the moment you shift topics, it forgets who it was pretending to be.

If voice matters – assign it each time! And when role is missing, the response can land soft and gray. Helpful, sure. But generic. Forgettable. No rhythm. No edge. No soul.

Ask something heartfelt, and you'll get something formal.
Ask something urgent, and you'll get a list.
Ask for help, and it will hesitate, not because it doesn't care –
but because you didn't show it how to speak.

Role doesn't just change the model's voice.

It changes its judgment, its values, its tone and its timing. It changes what gets left out. What gets softened. What gets spotlighted.

Without role, you get output.

With role, you get presence.

Use Any Role You Can Think Of

"Act as a [role]."
"Speak like a [persona]."
"Respond in the voice of a [type of person]."
"Imagine you are a [professional / identity / archetype]."
"Write this as if you are [personality or relationship role]."

Here are some examples:

"Act like a startup mentor giving tough love."
"Speak like a Gen Z TikToker hyping a product."
"Respond like a senior policy advisor under pressure."
"Write as if you're a parent talking to a worried child."
"Explain it like a stand-up comic who's also good at math."

These can go at the beginning or end of a prompt. Both work.

Do Role and Request Go in the Same Prompt?

Yes. Best practice is to assign Role within the same prompt as your Request.

Why? Because:

- It provides context in the moment
- It eliminates reliance on memory
- It ensures the model is aligned with that specific response

You can assign role at the top:

"Act like a kind but firm manager. Now write an email giving feedback on missed deadlines."

Or at the bottom:

"Give me 3 subject line options for this product launch. Respond like a creative director at an ad agency."

Will ChatGPT Remember the Role?

Sometimes, but not reliably, although GPT memory improves with each model release.

If you're in a single thread and the tone continues, it may loosely carry over. But role is fragile. It degrades over turns, especially if you change topics or reset tone.

If voice matters, restate the role.
Don't assume it'll remember. It won't persistently simulate a persona unless you remind it.

What Happens If You Skip the Role?

If you don't assign a Role, the model will respond with:

- Neutral tone

- Formal structure
- Slightly generic phrasing
- Mild assumptions based on common patterns

It's not wrong. But it's often bland, beige, and forgettable. Think default Wikipedia voice. Useful, but without eloquence.

In creative work, this makes it fall flat.
In business, it can sound vague or misaligned.
In emotional contexts, it might come off cold or robotic.

How Can Role Be Misused?

Roles can be used poorly. Here how:

1. Vague Roles

"Act like a good writer."

—is open-ended. What kind of good writer? Poetic? Technical? Journalistic?

2. Conflicting Roles and Requests

"Act like a strict lawyer. Now write a love letter."

Unless intentional for irony, this creates tone confusion.

3. Overly Prescriptive Roles

"Act like my therapist, but use no emotion, contractions, or metaphor."

Now you've handcuffed the model into an unnatural voice.

4. Inconsistent Role Use

"Sell this product. Act like a relaxed, helpful salesman."

then:

"Push the product aggressively."

Switching tone drastically without reassigning role or resetting voice.

The Letter That Didn't Land

Priya had just been promoted.

It had been a quiet climb — years of steady work, no drama, no spotlight.
But now, after five years, she was stepping into a team lead role.
Her first task: send a note to the team she'd just been asked to lead.

She opened ChatGPT and typed:

"Write an announcement email that says I've been promoted to team lead. Be polite, brief, and professional."

It delivered exactly what she asked for:

Subject: Leadership Transition

I'm pleased to share that I'll be stepping into the role of Team Lead effective Monday. I'm grateful for the opportunity and look forward to continuing our work together. Please feel free to reach out as we navigate this transition.

It was correct. Clean. Balanced. She read it twice. It felt… safe. She sent it.

No replies. No congratulations. No comments at all.

Just silence.

That afternoon, she overheard someone say, "I didn't even know it was her. Sounded like something HR wrote."

The truth? She hadn't asked the model to write like *her*. Or like a peer. Or even like a human with a voice. She'd asked for "brief and professional." And that's what she got. The model didn't fail. The prompt did. Not because of the words.

But because of the absence of a role.

Post-Game: The Voice That Wasn't Invited

Priya didn't do anything wrong.

She was polite. Professional. Appropriate. But those aren't the qualities that make people lean in. They're the ones that make people scroll past. Her note had no fingerprints on it. No trace of who she was. No rhythm, no warmth, no wink of familiarity.

She wrote like someone trying not to make a mistake. Which meant no one felt invited to celebrate with her.

And all it would've taken was this:

"Write an announcement email letting my team know I've been promoted to team lead. Keep it warm and humble. Speak like a colleague who's genuinely grateful – and maybe a little nervous. Add a sentence that shows I'm excited to support them, not just lead them."

That's not a fancy prompt. It's a clear one. It defines the who – not just the what.

And here's what the model might've returned:

Subject: I Get to Brag About You Now

Just a quick note to share that I've been asked to step into the team lead role starting Monday. I've been learning from all of you for the last five years, and I still am – so stepping into this new role feels a little surreal (and a little terrifying, let's be honest).

I'm genuinely excited to support this team – not just from a new seat, but as someone who still remembers what it felt like to join it. I want to keep what's working, and keep listening to what's not.

Thanks for the kind words (in advance or eventually). My door's open. Always has been.

That's voice. That's presence. Same news. Different impact.

Because the role she played in that prompt wasn't "new manager." It was human being with something to say.

And that's who people respond to.

The Takeaway

Role gives the model permission to adopt a voice. Without it, you're talking to a chorus with no soloist.

Use role when:

- Tone matters
- Authority matters
- Personality matters
- Relationship dynamics matter

And if in doubt: Just say who you want the model to be.
It'll do the rest – beautifully.

Chapter Reflection: Role

The model can mimic any voice – but only if you ask for one.

We've seen what happens when you give it the wrong tone, or no tone at all. Even good information can fall flat if the voice behind it feels distant, generic, or worse – inauthentic.

But with the right role – a mentor, a friend, a critic, a poet – the model speaks like it knows your audience.
Because it does. You showed it how.

Role isn't decoration. It's how you shape authority, intimacy, tone, posture, rhythm, restraint.

Sometimes it's bold. Sometimes it's subtle. But it always matters.

Because a request with no voice
is a story with no narrator.

91

Are There Any Limits?

You've told it what you want. You've shown it how it should sound.
You've given examples. You've provided background. You've chosen
the voice. But what about the edges? The boundaries? What's allowed?
What's off-limits?

That's where we go next: constraints.
The art of saying: "Yes – but only like this."

CHAPTER 6: WHAT ARE THE CONSTRAINTS?

Constraint is not limitation – it's leverage. It's the sharp edge that gives shape to infinite possibility. Want fewer words? Set a word count. Want only rhyming answers? Say so. Constraints make outputs smarter, tighter, and often far more creative.

This chapter shows how to rein in ChatGPT not to limit its ability, but to focus it:

"Summarize this article in three bullet points, no more than 12 words each. No adjectives."

ChatGPT might respond:

- Users abandoned app after price increase
- Competitor launched similar product last week
- Support team overwhelmed by refund requests

That's a boardroom slide, not a blog post. Clean, cold, direct – because you asked it to be. Invoke a younger, kinder voice:

> *"Rewrite this paragraph for a sixth-grade reading level. Keep it friendly."*

for a response like this:

> Sometimes, our brains get really busy. When that happens, it's hard to focus. Taking deep breaths or a short walk can help you feel better.

Not dumber. Just clearer. Constraint ≠ simplification. It's refinement:

> *"Write a welcome email in under 60 words. Use no exclamation marks. Make it warm without sounding salesy."*

might get you:

> Welcome aboard. We're glad to have you with us. If you ever need anything, just reach out. We're here to help – quietly and completely.

That's not just text. That's tone control. And word control:

> *"Generate five startup ideas in ten words or less. Avoid the words 'AI' or 'blockchain.'"*

could yield:

- Subscription snacks based on your DNA
- Digital wardrobe that plans outfits
- Kids' bank that teaches saving through stories
- App that rates your posture in real time
- Tool for recording family history, one voice message at a time

Constraints don't limit creativity. They guide it.
  Like banks on a river – they give the current its power.

Reveal What Matters

You don't constrain the model to make it smaller. You constrain it to make it more precise. More useful.
More like what you meant in the first place.

Because the real enemy of clarity isn't error – it's excess. Too many words. Too many ideas. Too many directions at once.

And the model – left alone – will happily give you all of it. Everything it can imagine, whether or not you need it. But when you add a constraint – a limit on length, tone, language, format, scope – you're not muzzling the response.

You're shaping it. You're saying: "No, not everything. Just this." And inside that boundary, something better appears: Something clear. Something tight. Something *intentional*.

This is where power lives. Not in asking for more.
But in asking for only what matters.

The Email That Spooked the Room

Jules ran comms at a venture-backed startup. Cool product. Strong quarter. They were prepping for a Series B. And now the founder wanted her to send a little "touch base" email to their biggest investor. Something casual. Light. Just to stay warm.

Jules was tired. She had five other things going. She opened ChatGPT and typed:

> *"Write a brief update email to our main investor about recent progress. Sound confident and friendly."*

She skimmed what came out. Looked fine. Sounded smart. She copy-pasted it, tweaked a word or two, and hit send.

Four minutes later, the founder was in her Slack:

"Did you tell them we're 'experimenting with alternate monetization frameworks'?"

Jules blinked.

"Did you mean to use the phrase 'stealth pivots'??"

The email hadn't been "fine." It had been way too much.

ChatGPT had filled in the gaps. It wanted to be helpful.
It took "progress" to mean "strategic evolution," and "confident" to mean "positioned for disruption."

The investor – sharp, old-school, allergic to nonsense – had replied with a single word: "Elaborate?"

And just like that, a simple touchpoint turned into a two-hour damage control call because no one told the model what *not* to say.

Post-Game: What Jules Should've Done

The prompt she gave was loose:

"Write a brief update email to our main investor about recent progress. Sound confident and friendly."

But "brief" is subjective. "Progress" is vague. "Confident" can spill into arrogant, evasive, or even risky. And with no constraints, the model tried to fill the silence with swagger. Here's what Jules *could* have written:

"Draft a short email – no more than 80 words – updating our lead investor on user growth and the new onboarding feature. Keep it specific and clear. Avoid buzzwords, changes of course, and finances."

Now the model knows:

- What to focus on
- What to ignore
- How long to speak
- How not to sound

It doesn't have to guess what "professional" means. It doesn't reach for filler. It doesn't invent drama.

It just writes what matters. Here's what that version might've looked like:

Hi Chris – just a quick update: The onboarding flow is live and already showing a 12% lift in activation. MAUs are up 8% month-over-month. No major surprises, just steady traction. Hope all's well on your end.

That's it. No hype. No pivot. No stealth. Just facts in clean shoes. And that would've been the end of it.

No Slack ping. No damage control. No explaining "alternate monetization frameworks" at 6:30 p.m. on a Thursday.

Constraint Types

A quiet tightening. A necessary narrowing. Not less – just clearer.

The model, left unchecked, will always try to give you everything. It will pile on options, offer extra lines,give you metaphors, summaries, analogies, hashtags, disclaimers, sometimes even quotes from people you didn't ask for. Not because it's wrong, but because it doesn't know where the edges are.

Constraints are how you draw those edges.

Not to shrink the response, but to give it form. Not to limit what the model says, but to control how much, how strongly, and in what direction. Here are the kinds of limits that matter most:

1. Length

Some truths only need a sentence. Others, a single word. If you don't say how much, the model might say too much.

When you set a length – by words, lines, bullets, paragraphs – you force the model to choose.
To cut. To condense. To commit.

It's not about brevity for its own sake. It's about pressure, the kind that makes language shine:

> *"Give me three bullets only."*
> *"Summarize this in a single sentence."*

"No more than 100 words. Keep it tight."
"One paragraph, not two."

The shorter the shape, the sharper the thought.

2. Tone

The model doesn't have a voice until you give it one. You don't just want the right information. You want the right feeling in the room. Tone constraints don't control what's said. They control how it's heard.

Kind. Casual. Wry. Reverent. Neutral but not cold. Excited but not obnoxious.

These aren't decorations. They're anchors:

"Make it friendly, but not unprofessional."
"Use a gentle tone – think therapist, not teacher."
"Make it sound like I'm explaining this to a friend, but not angrily."
"Respectful. Not apologetic. Firm."

Tone can make the difference between read and received.

3. Vocabulary

You can say something ten different ways, but the words you choose decide who understands it.

Vocabulary constraints shape the accessibility of your response. They tell the model how complex to be, how dense, how plain.

Not all clarity is simple, but simplicity often helps. You can ask for common words. Or for domain-specific ones. Or for the kind of language a 12-year-old could love. Here are some ideas:

> *"Use only the most common words."*
> *"Avoid jargon – speak plainly."*
> *"Use precise legal terms where appropriate."*
> *"Explain this as if I'm new to the field and avoid involved terminology."*

It's not about dumbing down. It's about letting people in.

4. Scope

The model knows a little about everything. But that doesn't mean you want all of it.

Scope constraints tell it where to look – and where not to wander. They don't change the answer. They narrow the lens.

You might limit it to a region. Or a moment in time. Or a point of view.

You're not closing doors. You're saying: Just this room. Not the whole house.

> *"Focus on the financial side of this."*
> *"Keep your answer within the 20th century."*
> *"Use examples only from biology."*
> *"Answer this like its a debate in a classroom."*

Scope is how you stay on track and keep the model from taking scenic detours.

5. Exclusion

Sometimes what matters most is what doesn't belong. Exclusion constraints tell the model what to leave out. Not just topics. But phrases. Clichés. Brands. Spoilers. Warnings you've already covered.

They're small but powerful – the guardrails that prevent a wrong turn in an otherwise right direction.

"Do not mention any competitors."
"Avoid buzzwords like 'cutting-edge.'"
"Leave out your usual safety disclaimers."
"Do not summarize the plot."

This is how you keep the voice clean. The content focused. The message intact.

> > >

These constraints – length, tone, vocabulary, scope, exclusion – are not limitations. They're *invitations to clarity.*

Each one says: Now that I've asked you what I want, here's how I want it. Not everything. Just this. And the model – grateful for the guidance – will finally stop guessing and start delivering.

Speaking of guidance, it's time for a parenting allegory.

The Mother and the Messy Room

The mother stood in the doorway of her daughter's room. She didn't cross the threshold. She didn't raise her voice. She just said:

"I need you to clean this up."

The daughter, who was twelve and dreaming about the collapse of galaxies, nodded vaguely and returned to sketching something furious in her notebook. Two hours passed. The room looked the same. Maybe worse.

The mother sighed, came back, said it again:

"I asked you to clean."

The daughter blinked. "I did. I put the clothes on the chair." The mother looked at the chair. It had collapsed under the weight of a half-laundry mountain. The floor was a terrain of paper, socks, snack fossils.

The mother was tired. The daughter was confused. They were both angry and neither understood why.

What the mother didn't say:

"Here's what I expect you to clean and not to clean."
"Don't just put a couple of things away and think the job is done."
"Clothes do not belong on the chair. Put them in the basket or in your bureau."

The mother made a request but gave no constraints. Models, like children, are apt to daydream without some discipline.

We say to the model: Now that I've asked you what I want, here's how I want it. Not everything. Just this. And the model will stop guessing and start delivering.

Constraints are the final element in prompt anatomy! Let's sum up this little journey.

CHAPTER 7: HOW TO PROMPT POWERFULLY

There's a moment after the map is drawn, after the training is done, after the roles are chosen, and the rules are laid down.

A quiet moment.

And in that moment, someone opens a blank chat window, types a single line, and waits. This chapter is for that moment.

You already know the parts:

- The Request: what you want
- The Examples: the pattern to follow
- The Context: where you're standing when you ask
- The Format: the shape you need the answer to take
- The Role: who's speaking and to whom
- The Constraints: what should be held back

Each chapter has taught you one. Each story showed how it works. But now it's time to see how they work together.

Elena didn't know she was teaching you request when she reached out to ask for help with a product launch, a poem, a life. She asked, again and again, until the shape of the ask sharpened into a tool.

The king didn't know his mapmaker was teaching format. He just wanted direction. But in showing him form, he gave him more than direction – he gave him vision.

The coach taught by doing. She didn't explain what an example was. She just said: "Like this.", "Try it that way.", "Let me show you what I mean."

And the little boys? They gave the model voice after voice until it laughed back at them – a pirate, a poet, a robot, a sister. They taught you role without knowing they were doing it.

Jules taught you constraints – the hard way. When no one said what not to say, the model filled the silence with smoke.

Jake, Priya, Nico... All of them were speaking with power.
Some knew it. Some didn't.
But you know it now.

Because this is the truth of it:

A prompt is not a question. It is a composition.

Like a score for an invisible orchestra. Like stage directions for a voice with no body. Like a letter sent to the future.

When you prompt powerfully, you are not "using AI."
You are crafting language that shapes response.

You are not waiting to be impressed.
You are directing the impression.

There is no perfect prompt. There is only the one that fits the
moment.
Tight where it needs to be. Open where it dares to be.

Some prompts are short. Some are careful.
Some are playful, strange, poetic, brave.
But the best ones – the powerful ones – are *composed.*

They don't just include a request. They show what the request means.
They don't just assign a role. They know who that role is speaking to.
They don't just limit. They leave room.

Every element influences the others.

A format can suggest a tone.
A constraint can force clarity.
An example can replace a paragraph of explanation.
A role can collapse the need for length.
A context can light your set.

These aren't tools in a box. They're strings on the same instrument.

And you –
you've been learning to play.

The Compleat Prompt

Let's see it all,
gently, in one ask:

"Act as a seasoned career coach. Based on the attached resume, write a short email this candidate Evan Pratt can send to a recruiter Gayle Marquez at a creative agency. Keep it under 100 words. Make the tone curious and confident. Don't include a subject line. Here's a sample of how Evan writes."

There it is:

- **Role**: a seasoned coach
- **Context**: a resume is attached
- **Request**: write an email for a job search
- **Format**: formal but friendly introductory email
- **Constraints**: under 100 words, no subject line
- **Examples**: a sample writing style is provided

All in one breath. Nothing wasted. Everything directed. The result?

Will feel like it was written by someone who knows exactly what they want – and how to ask for it:

Hi Ms. Marquez,

I came across your agency's work on the Vega campaign. Beautifully done. I'm exploring new creative opportunities where voice and brand intersect and your team's approach really stood out. Would love to connect and hear what you're building next.

Warmly,
Evan Pratt

Just under 70 words. No filler. Curious, but not overeager. Confident, but not performative.

No buzzwords. No backstory. Just enough for the door to open. Because the prompt didn't just say what it wanted.

It made the path *walkable.*

The Final Prompt

"Act as a poetic and practical author.

The reader is finishing a book about how to write powerful prompts for large language models.

Based upon the book's teachings, write a final passage that will feel to the reader like they are closing a guide, not to using a tool, but to one of the most important skills in life:

Making an effective request.

Use the same voice as the book – simple, lyrical, confident. Limit to 120 words.

Do not summarize the book. Do not repeat the chapter titles.

Speak to the reader as someone who now understands the structure of a prompt as a way of structuring their communication, their instructions, and their questions.

Leave them with something that feels quiet and true."

And the model – having heard everything – replies:

You know how to ask now.

Not just questions,
but *the kind that opens doors.*

You know what to include.
What to leave out.
How to make the words fit the moment.

So go ask.

Not always for answers.
But for understanding.
For clarity.
For connection.

Let the silence before the prompt
become part of the message.
Let the way you ask
reflect who you are.

And remember –
every great answer
was once
a better
question.

This book ends.
The conversation begins.

APPENDIX A: 500 POWER PROMPTS

Here are 500 power prompts arranged in our categories of request, examples, format, context, role, and constraints.

Prompts that end with a colon (":") expect you to append or upload your own content as part of the prompt. For example:

"Make this sentence more gentle: How many times have I told you to close that door?"

Here are your prompts beginning with basic requests.

Request

Here are 100 prompt request examples in 10 themed subcategories.

1. WRITING & EDITING

Rewrite, rephrase, and polish like a pro editor.

Rewrite for Tone
How can I make this text sound more assertive?

Clarity Check
Identify all unclear parts of this paragraph.

Style Mimicry
How would Hemingway describe a fishing trip in the mountains?

Jargon Translation
Simplify this technical document for a 6th grade reading level.

Elevator Pitch
Create a 30-second pitch from this content.

Summarize Cleanly
What are the five key takeaways from the latest climate report?

Concise Rewrite
How can I condense the text of the US Constitution while preserving its core message?

Simplify Language
Convert this text to plain English.

Tone Shift
How would this sound with more empathy?

Add Humor
Add light humor to this text without being silly.

2. CONTENT CREATION

Turn rough ideas into compelling content across multiple formats.

Blog to Thread
Convert this blog post into a tweet thread.

Headline Variants
What are 10 alternative headlines for this idea?

Social Adaptation
Transform this content into a LinkedIn post.

Script Starter
How would this article work as a short video script?

Lead Magnet
What free resource could I create from this content?

Newsletter Opener
Draft an engaging intro paragraph for this newsletter.

Call to Action
Craft a call-to-action for an email newsletter about sustainable fashion.

SEO Juice
Enhance this article with relevant keywords.

Punchy Rewrite
Rewrite this to grab attention immediately.

Visual Prompt
Describe an image that would complement this post.

3. COMMUNICATION

Sharpen your message whether you're pitching, persuading, or politely declining.

Polite Rejection
Write a kind but firm rejection email.

Positive Feedback
What encouraging feedback would you give on this work?

Client Update
Create a project update email from these notes.

Status Request
Draft a polite request for a status update.

Upsell Pitch
How would you pitch an upgrade to an existing customer?

Warm Follow-up
Write a follow-up for someone who hasn't responded.

Conflict Resolver
Craft a diplomatic response to this complaint.

Team Praise
Write a congratulatory message for a team member who completed a challenging project.

Thank You Note
Write a short, sincere thank-you email.

Cold Outreach

Introduce yourself to a potential collaborator.

4. TRANSFORMATION

Change format, perspective, or medium while keeping meaning intact.

Listify This

Transform this paragraph into a bulleted list.

Turn into Story

Can you rewrite this summary as a narrative?

Poemify

Create a short poem from this text.

Into a Tweet

Condense this into a 280-character tweet.

As Dialogue

Convert the concept of artificial intelligence into a dialogue between a teacher and student.

Make It a Metaphor

What metaphor best explains this concept?

Flip the Perspective

How would the story of Little Red Riding Hood look from the wolf's perspective?

As a Riddle

Transform the concept of photosynthesis into an engaging riddle.

Slide Format

Convert this content into presentation slide bullets.

Table-ize

Create a two-column comparison table from this information.

5. EXPLANATION

Clarify, demystify, and break things down step-by-step.

Explain Like I'm 5

How would you explain this to a 5-year-old?

Why It Matters

Why is this concept important in everyday terms?

How It Works

Can you explain exactly how a car functions from start to finish?

What's the Catch?

What are the potential drawbacks of this idea?

Compare to a Car

Explain this using a car analogy.

Draw Me a Timeline

Organize this information chronologically.

Show Me the Math

Could you break down the E=mc squared equation step-by-step with clear explanations?

Use an Analogy

What creative comparison would help explain this concept?

From the Top

Explain this as if I have no prior knowledge.

Summarize and Expand

What's the core message of the IRS, and how would you elaborate on it?

6. STYLE EXPANSION

Shift voice, tone, and energy by cloning style.

Make It Poetic

Rewrite this like a poem using the same idea:

Bolder Version

Rewrite this to be more provocative:

More Playful

Keep the structure but add playfulness:

More Corporate

Turn this into business-speak with buzzwords:

From Academic to Accessible

Rewrite this so a general audience can understand:

Spice the Voice

Make the tone more colorful and expressive:

Flatten the Jargon

Rewrite this so it's understandable by a layperson:

Snark Inject

Add subtle sarcasm while keeping it intelligent:

Charming Rewrite

Rewrite this to be warm and endearing:

Subtle Sales Pitch

Add a persuasive undertone without being obvious:

7. VISUAL EXAMPLES

Base the prompt on an image, layout, or formatting clue.

Describe This Image

Write a caption for this visual:

Emulate This Chart

Here's a table – make another that matches the format:

Design Prompt

Based on this layout, describe a matching design:

Diagram It

Convert this idea into a diagram with components:

Create Matching Icons

Based on this list, describe an icon for each:

Interface Copy

Write interface labels like the example shown:

Flowchart Language

Turn this explanation into a flowchart format:

Side-by-Side Copy

Match the layout of this content in a new topic:

Describe a New Scene

Match the tone and vibe of this photo description:

Match a Slide

Write bullet points for a slide like the one shown:

8. ROLE-BASED EXAMPLES

Use personas or archetypes as demonstration tools.

Therapist Version

Respond to this like a therapist:

CEO Rewrite

Say this as a tech CEO would:

Student Voice

Say this from the point of view of a college student:

Chef's Take

Describe this process like a professional chef:

Professor Rewrite

Deliver this answer like a philosophy professor:

Startup Founder Spin

Pitch this like a founder to investors:

Customer Service Style

Turn this into a polite helpdesk response:

Comedian Rewrite

Tell this like a standup comic would:

Coach Version
Motivate like a personal trainer.

News Anchor Version
Deliver this line like a breaking news segment:

9. INTERACTIVE EXAMPLES

Use dialogue, quiz, or engagement structure in prompt.

Call and Response
Based on this call, write the next response:

Quiz Me
Build a quiz that mimics this format:

Dialogue Prompting
Continue this back-and-forth exchange:

Correction Mode
Given this answer, write the tutor's corrections:

Feedback Loop
Write an example, then write feedback on it:

Turn and Respond
Respond to this statement as if you're in the room:

Q&A Pattern
Follow this format of question/answer:

Game Style Reply
Match the tone of this text-based adventure:

Mimic This Format

Repeat this interview Q&A format with new topics:

Ping-Pong Format

Go back and forth, alternating tone with each line:

10. META EXAMPLES

Use examples that model prompting itself.

Prompt Anatomy

Break down this prompt into its components:

Prompt and Output Pair

Give me both the prompt and a great response:

Bad Prompt Fix

Here's a weak prompt – improve it and show why:

Prompt Variation

Write five prompts that produce similar outputs:

Prompt Expansion

Take this and add examples, role, and format:

Turn Prompt to Template

Rewrite this as a reusable prompt template:

Add Constraints to Prompt

Modify this to include a word limit and tone:

Prompt Tuner

Change this prompt to get more creative answers:

Prompt Formatter

Rewrite this in a structure that's easy to copy/paste:

Meta Mirror

Analyze this example and rewrite it to teach prompting:

Examples

Here are 100 examples prompts in 10 themed subcategories.

1. PATTERN MATCHING

Teach the model to mimic your structure and flow.

Follow the Form
   Here's one bio – generate another with the same structure:

Mimic Style
   Copy this sentence's tone and cadence without repeating words:

Repeat the Format
   Generate three examples in this format:

Fill in the Blank
   Continue this structure logically:

Match this Outline
   Use this outline and fill it in with new content:

Repeat with Variation
   Produce five variations based on this input:

Finish the Set
   Here are four list items – add three more that fit:

Imitate the Opening
   Write a paragraph starting just like this one:

Reflect this Rhythm
   Write a response with matching sentence length and flow:

Structure Swap

Use the same format but change the topic completely:

2. CONTRAST & CORRECTION

Show bad examples, flawed drafts, and then improve them.

Improve the Draft

Here's a weak version. Rewrite it to be stronger:

Spot the Flaws

Critique this paragraph and suggest changes:

Polish the Pitch

Here's a rough idea – make it sound compelling:

Fix the Mistakes

Identify and correct grammar issues in this text:

Upgrade the Argument

Strengthen this reasoning with better logic:

Refine the Reply

This email sounds passive. Make it more confident:

Before and After

Give me the "bad" version and the improved one:

Rewrite to Respect

This sounds rude. Rewrite it with kindness:

Fix the Flow

This feels disjointed – make it smooth:

Reverse the Tone
Rewrite this optimistic note to be somber:

3. COMPARATIVE LEARNING

Two styles, two outcomes – highlight the difference.

Option A vs B
Compare these two summaries and explain what changed:

Vary the Ending
Keep the beginning – change how it ends:

Dual Rewrite
Rewrite this both formally and casually:

Split the Voice
One version should sound human, one robotic:

Subjective/Objective
Write two takes – one personal, one analytical:

Rewrite in Two Genres
One version sci-fi, the other romance:

Tone Flip
First version skeptical, second enthusiastic:

Change the Lens
Describe this scene once as a child, once as a detective:

Reframe the Hook
One version as a mystery, one as comedy:

Two Audiences

One for beginners, one for experts:

4. PROMPTED BY EXAMPLE

Prompt the model with a prototype to follow.

Write More Like This

Based on this, write three more in the same tone:

Extend the Pattern

Use this example and continue the list:

Rephrase to Match

Rewrite this sentence to match the given one's rhythm:

Style Match

Write in the style of this paragraph:

Sentence Twin

Create a new sentence that mirrors this one's structure:

Copy the Syntax

Build a new example with the same grammar flow:

Mirror the Function

The purpose of this message is to inspire. Write another:

Match the Mood

This sample is tense. Write another with the same mood:

Parallel Rewrite

This sounds good. Write a new one with a new topic:

Follow the Example

Here's a Q&A pair – give me three more like it:

5. STEP-BY-STEP BY EXAMPLE

Use examples that demonstrate a process or sequence to follow.

Email Correction Pattern

Here's how I would fix this email: "Original: 'I need the report.' Fix: 'I need the quarterly sales report by Friday at 3pm. [Added specificity about which report and deadline]'" Apply this correction pattern to improve this vague request:

Tutorial Structure Example

Notice this tutorial format: "Step 1: Identify the problem (The login button doesn't work). Step 2: Check for errors (Console shows 404 error)." Create a similar step-by-step analysis for troubleshooting a slow website.

Annotation Style

Look at this annotation example: "Original text: 'We should meet.' Annotation: [Too vague - specify time, place, and purpose]" Use this annotation style on the following sentence:

Comparison Framework

Here's how I compare two options: "Option A: High cost, low maintenance | Option B: Low cost, high maintenance" Create a similar comparison for these two software solutions:

Document Format Example

This is how I format a project brief: "PROJECT: [Name] | DEADLINE: [Date] | OBJECTIVES: [Bullet points]" Format the following information using this template:

Template Pattern

Study this template: "CUSTOMER: [Name] | ISSUE: [Problem] | RESOLUTION: [Solution] | FOLLOW-UP: [Next steps]" Create a blank version with appropriate placeholders for a product return process:

Two-Phase Analysis Example

Notice this approach: "Summary: The market is growing 5% annually. Expansion: This growth is driven by three factors..." Apply this summary-then-expand pattern to analyze this trend:

Demonstration Style

Here's how I explain while doing: "I'm selecting the text [selection shown], then clicking Format -> Bold [screenshot]." Use this show-and-tell approach to explain how to create a chart:

Result-First Pattern

See this structure: "Final result: A clean database. Process to get there: 1) Remove duplicates..." Use this result-then-process pattern to explain achieving better team communication:

Line-by-Line Editing Example

Study this editing approach: "Original: 'The meeting was good.' Edit: 'The quarterly review revealed promising Q3 results.' [Added specificity and positive outcomes]" Apply this detailed editing style to improve this sentence:

6. STYLE EXPANSION

Shift voice, tone, and energy by cloning style.

Make It Poetic

Rewrite this like this poem using the same idea:

Bolder Version

Rewrite this to be more provocative like this other note:

More Playful

Keep the structure but add playfulness:

More Corporate

Turn this into business-speak with buzzwords like in the attached product description:

From Academic to Accessible

Rewrite this summary so a general audience can understand it like this other summary:

Spice the Voice

Make the tone more colorful and expressive like this other note:

Flatten the Jargon

Rewrite this so it's understandable by a layperson similar to the attached explanation:

Snark Inject

Use sarcasm like in this note to say thank you:

Charming Rewrite

Write a warm and endearing thank you like this one:

Subtle Sales Pitch

Use the same undertone in this example:

7. VISUAL EXAMPLES

Base the prompt on an image, layout, or formatting clue.

Describe This Image
Write a caption for this visual:

Emulate This Chart
Here's a table – make another that matches the format:

Design Prompt
Based on this layout, describe a matching design:

Diagram It
Convert this idea into a diagram with components:

Create Matching Icons
Based on this list, describe an icon for each:

Interface Copy
Write interface labels like the example shown:

Flowchart Language
Turn this explanation into a flowchart format:

Side-by-Side Copy
Match the layout of this content in a new topic:

Describe a New Scene
Match the tone and vibe of this photo description:

Match a Slide
Write bullet points for a slide like the one shown:

8. ROLE-BASED EXAMPLES

Use examples to demonstrate specific role patterns and voices.

Therapist Pattern

Here's how a therapist might respond: "I hear that you're feeling overwhelmed. What specific aspects feel most challenging right now?" Write a similar response about anxiety:

CEO Communication Pattern

Notice how this CEO communicates: "We're not just building features – we're solving industry-wide problems at scale." Use this pattern to describe a new product launch:

Student Writing Style

This is how a college student writes: "The professor's lecture was interesting but I'm still confused about the main concept. Anyone have good notes?" Write something about a group project in this style:

Chef's Descriptive Pattern

Follow this chef's descriptive style: "We're gently coaxing the flavors out, allowing the aromatics to bloom without browning." Describe making coffee using this pattern:

Professor's Explanation Format

Here's a philosophy professor's explanation pattern: "If we examine this concept dialectically, we find an inherent tension between two principles..." Continue this explanation about freedom:

Startup Pitch Example

Study this pitch pattern: "We've identified a $2B market gap where existing solutions fall short on three key metrics." Create a similar pitch for an imaginary app:

Customer Service Template

Follow this helpdesk response example: "I understand how frustrating this issue must be. Let's get this resolved in three simple steps." Write a response about a delivery problem:

Comedian's Rhythm

Notice this comedy pattern: "So I tried meditation – turns out my thoughts are even worse when I'm paying attention to them." Create a similar joke about technology:

Coach's Motivation Formula

This is how a coach motivates: "You're not tired – you're just redefining what you're capable of. Three more reps!" Write a similar motivational message about learning:

News Anchor Structure

Follow this news delivery pattern: "Breaking tonight – developments in the story we've been following. Officials now confirm three key details." Create a similar news intro about a weather event:

9. INTERACTIVE EXAMPLES

Use dialogue, quiz, or engagement structure in prompt.

Call and Response

Based on this call, write the next response:

Quiz Me

Build a quiz that mimics this format:

Dialogue Prompting

Continue this back-and-forth exchange:

Correction Mode

Given this answer, write the tutor's corrections:

Feedback Loop

Write an example, then write feedback on it:

Turn and Respond

Respond to this statement as if you're in the room:

Q&A Pattern

Follow this format of question/answer:

Game Style Reply

Match the tone of this text-based adventure:

Mimic This Format

Repeat this interview Q&A format with new topics:

Ping-Pong Format

Go back and forth, alternating tone with each line:

10. META EXAMPLES

Use examples that model prompting itself.

Prompt Anatomy

Break down this prompt into its components:

Prompt and Output Pair

Give me both the prompt and a great response:

Bad Prompt Fix

Here's a weak prompt – improve it and show why:

Prompt Variation

Write five prompts that produce similar outputs:

Prompt Expansion

Take this and add examples, role, and format:

Turn Prompt to Template

Rewrite this as a reusable prompt template:

Add Constraints to Prompt

Modify this to include a word limit and tone:

Prompt Tuner

Change this prompt to get more creative answers:

Prompt Formatter

Rewrite this in a structure that's easy to copy/paste:

Meta Mirror

Analyze this example and rewrite it to teach prompting:

Format

Below are 100 prompts grouped into 10 formatting styles. Use them to bend structure to your will.

1. LISTS & BULLETS

Use bullets to shape clarity, rhythm, and hierarchy.

Top Ten List
Create a top 10 list with brief descriptions.

Bullet Point Summary
Can you summarize this into themed bullet points?

Pros and Cons
Format this as a two-column pros and cons list.

Checklist Format
Convert this outline into a to-do checklist.

Quick Wins
What are 5 easy actions to take immediately?

Steps in Order
Create a numbered list for completing this task.

Ranking List
How would you prioritize the key factors when choosing a career path?

Alternatives List
Can you suggest 5 different approaches to attaining world peace?

Grouped Categories

Organize this information into categorized bullet points.

Bolded Headings

Format this with bolded titles for each bullet group.

2. TABLES & MATRICES

Compare and clarify with rows and columns.

Comparison Table

Build a 3-column table comparing these features.

Pros, Cons, Use Cases

Create a table showing pros, cons, and use cases.

SWOT Table

Format this information as a SWOT matrix.

Risk Matrix

How would you chart likelihood vs. impact for these risks?

Pricing Table

Design a pricing plan comparison table.

Timeline Table

Create a timeline table with dates and milestones.

Team Roles Table

Organize these team roles and responsibilities in table format.

Multi-Choice Grid

Build a table of choices with scores in each column.

Editable Table Format

Format this as a markdown table for easy copying.

Table with Notes

Create a comprehensive table of the best Roombas with an additional column for helpful notes.

3. OUTLINES & STRUCTURES

Direct flow, sequence, and hierarchy with outlines.

Basic Outline

Organize this idea into a three-level outline:

Presentation Structure

Format this as a 5-slide presentation:

Narrative Arc

Turn this into a beginning, middle, and end:

Essay Framework

Structure this into intro, argument, counterpoint, and conclusion:

Process Steps

Outline the full workflow for this task:

Hierarchy Tree

Display this as a nested structure:

Course Outline

Build a syllabus-style layout for this topic:

Workshop Flow

Plan a workshop session with time blocks:

Script Breakdown

Break this into scenes or chapters:

Milestone Plan

List key deliverables with dates and owners:

4. SLIDE & DECK FORMATS

Convert ideas into presentations and slide-friendly formats.

Slide Bullets

Convert this into bullets for a single slide:

Slide Titles

Create 5 strong titles for a presentation deck:

Slide Order

Rearrange this info into a compelling slide order:

One-Slide Summary

Reduce this to one powerful summary slide:

Problem–Solution Slide

Build a two-part slide: challenge and fix:

Pitch Deck Format

Outline a 10-slide startup pitch:

Headline + Bullet

For each idea, generate a slide title and 3 bullets:

Big Idea Slide

Frame this idea as a single bold slide:

Side-by-Side Comparison

Format this for a two-column slide:

Final Slide CTA

Write a compelling closing slide with a call to action:

5. CODE & LOGIC STRUCTURES

Shape outputs with precision for devs and tech thinkers.

JSON Format

Structure this content as a JSON object:

Python Dict

Format these variables in Python dictionary style:

If–Then Logic

Turn this into a list of if/then statements:

Decision Tree Format

Structure this logic as a branching decision tree:

API Request Format

Create a sample API request with headers and parameters:

Markdown Format

Convert this into clean markdown syntax:

Data Table

Output this info as a CSV-style table:

Regex Examples

Generate regex patterns based on this text:

YAML Format

Write this structured prompt as YAML:

Code Comment Format

Annotate this code block with inline comments:

6. QUESTION & ANSWER FORMATS

Format as interviews, FAQs, and conversational flows.

FAQ Style

Turn this into a frequently asked questions list:

Interview Format

Write this as a Q&A interview:

Quiz Format

Structure this content as a multiple-choice quiz:

Socratic Dialogue

Create a question-driven conversation:

Rapid Fire Q&A

Make this a quick back-and-forth sequence:

Explainer with Questions

Interweave questions and answers:

Three-Tiered Answer

Provide a short, medium, and long answer:

Inquisitive Prompt

Write a question designed to provoke deep insight:

Structured Inquiry

Use numbered questions to guide learning:

Conversational Coach

Format this as a guided dialogue with responses:

7. EMAIL & MESSAGE STYLES

Shape output as professional or casual messages.

Formal Email

Write this as a polished business email:

Casual Message

Turn this into a friendly Slack message:

Follow-up Email

Write a polite follow-up on this topic:

Thank You Email

Format a warm and professional thank-you:

Customer Support Email

Turn this into a support response email:

Cold Outreach

Write an email introducing yourself to a stranger:

Apology Email

Craft a sincere and effective apology message:

Sales Email

Format this into a persuasive cold pitch:

Internal Update

Convert this into a team status update:

Auto-Reply Message

Write an out-of-office or away message:

8. WEB & CONTENT LAYOUTS

Turn ideas into web-friendly blocks.

Landing Page Format

Build a landing page layout from this copy:

Hero + CTA

Write a hero headline and subhead with CTA:

Blog Post Sections

Break this into H2/H3 structure for a post:

Sidebar Format

Reformat key ideas as sidebar callouts:

Card Layout

Describe this idea in 3 stacked cards:

Feature Section

Highlight this product's features with icons and bullets:

Pricing Page Copy

Write copy for 3 pricing tiers with benefits:

Callout Boxes

Turn these tips into callout blocks:

Grid Format

Layout this idea as a 3x3 grid of content:

Page Wireframe

Describe a full-page content structure:

9. SCRIPTS & PERFORMANCE FORMATS

Build scenes, speeches, and roleplay instructions.

Script Format

Write this as a character script:

Two-Person Dialogue

Turn this into a back-and-forth scene:

Speech Format

Format this as a short speech:

Monologue Style

Express this as an internal monologue:

Narration Block

Write this as voiceover narration:

Stage Directions

Add scene directions and cues:

Interactive Script

Format this like a choose-your-own-path interaction:

Performance Pitch

Write a pitch with stage presence:

Podcast Script

Convert this idea into a podcast outline:

Host + Guest Format

Create a segment for an interview show:

10. MIXED MEDIA FORMATS

Go cross-modal: text plus visuals, audio, and interaction.

Image Prompt Format

Write a prompt for generating this image:

Video Description

Describe the shot list for this video idea:

Audio Script

Create an audio ad script with timestamps:

Interactive Layout

Design an interactive page structure:

Visual Storyboard

Build a 4-panel storyboard description:

AR/VR Prompt

Write an immersive prompt for spatial experience:

Story + Sound

Combine a scene with suggested soundtrack:

Printable Format

Turn this into a printable worksheet:

Poster Copy

Write copy for a striking poster layout:

Mixed Format Prompt

Combine text, image, and interaction in one prompt:

Context

Below are 100 context prompts grouped into 10 subcategories.

You will encounter prompts that are meant to be used in combination with other prompts. Here are three prompts for example:

> *"Tell me why I should care about nutrition."*
> *"Make it sound like my gym coach."*
> *"Don't say anything bad about beer or pizza."*

These can be used all at once or separately depending upon the conversation you wish to have with ChatGPT.

1. PURPOSE FRAMING

Clarify the goal behind the prompt with specific context:

Summarize to Inform
> *A busy executive needs to decide whether to invest in a new software platform. They have 2 minutes to review your summary before their next meeting.*
>
> *What key decision factors from these quarterly results should be highlighted?*

Convert to Persuade
> *Your audience consists of experienced engineers who have tried similar solutions before and were disappointed. They value data over promises.*
>
> *How can this technical description address their specific concerns?*

Explain to Teach

Your audience is a group of marketing interns with no prior experience in data analytics. They need to understand this concept to complete their first project.

Create a beginner-friendly lesson from these advanced analytics principles.

Inspire Action

Your team has been working remotely for months and motivation is declining. They need a clear reason to invest energy in this new initiative.

Energize your team with this project announcement.

Report with Authority

You're presenting these findings to the board of directors who will use this information to make a strategic decision. They respect expertise and clarity.

Present these research findings with implications for the five-year plan.

Persuade Gently

Your client is hesitant about increasing their budget but needs the additional services. They have been with your company for five years.

Rewrite this pricing update to maintain the relationship while softly conveying the value of the increased investment:

Encourage Exploration

You're launching a new community platform where user

participation is essential. First-time visitors need to feel comfortable contributing.

Transform this community guidelines document into an inviting introduction that encourages new members to explore and engage:

Frame as Advice

A junior colleague has asked for your guidance on a challenging project similar to one you completed last year. They respect your experience but want to develop their own approach.

What guidance would empower them to make their own decisions?

Reduce Resistance

Your department needs to adopt a new workflow that will initially require extra effort but will save time long-term. Several team members have expressed concerns.

How would you present this process change to acknowledge concerns while emphasizing benefits?

Pitch Softly

You're emailing a potential client who requested information but explicitly mentioned they dislike aggressive sales tactics. They're comparing several options.

Draft a non-pushy follow-up that presents your solution's value.

2. AUDIENCE TARGETING

Shape the tone and language for the right listener with specific audience context.

Write for Teens

Your audience is a group of 15–17 year olds who primarily consume content on TikTok and Instagram. They're tech-savvy but have short attention spans and are allergic to anything that feels condescending or inauthentic.

Rephrase these cybersecurity guidelines to engage and resonate with this teenage audience:

Make it Executive-Friendly

You're communicating with a CEO who receives over 300 emails daily and has 15-minute meeting blocks. They value data-backed insights and strategic implications over technical details.

Adapt this technical product update for this time-constrained C-level executive:

Explain to My Grandma

Your grandmother is 78, uses a smartphone for calls and photos only, and has never used a streaming service. She's intelligent but unfamiliar with digital technology terminology.

Make an explanation of cloud storage accessible to her without being patronizing.

Technical Peer Mode

You're addressing a room of senior software engineers who have been working with this technology stack for 5+ years. They respect technical precision and evidence-based arguments.

Write this architectural proposal for this expert audience:

Investor Audience

You're presenting to venture capitalists who focus on early-stage

B2B SaaS companies. They've seen hundreds of pitches and are primarily concerned with market size, competitive advantage, and unit economics.

Shape this product description as a compelling pitch to these experienced investors:

Team Briefing Format

You're updating a cross-functional team that includes designers, developers, and marketers. Some team members are remote, and everyone is juggling multiple priorities.

Write this project update in a way that respects their time while ensuring everyone understands their next actions:

Customer-Facing Tone

Your customers are enterprise procurement professionals who need clear, factual information without marketing hype. They're evaluating multiple vendors and need to justify decisions to their stakeholders.

Reword this product announcement for this external business audience:

Young Reader Voice

You're creating educational content for 11-13 year olds who are learning about climate science. They have basic scientific knowledge but are still developing abstract thinking skills.

Reframe an explanation of carbon capture for middle school comprehension.

Informed But Busy

Your audience consists of healthcare professionals who understand medical terminology but are reading this between patient appointments. They need to extract key information in under 60 seconds.

Communicate these research findings quickly and clearly for these time-pressed experts:

Reader is Suspicious

You're addressing consumers who have been misled by similar companies in the past. They're researching alternatives and approaching all claims with healthy skepticism.

Write a product description for readers who don't automatically trust your claims.

3. EMOTIONAL TONE

Set the emotional backdrop for the output with specific emotional context:

Make It Uplifting

Your audience has just experienced a difficult quarter with budget cuts and team restructuring. They need to see a path forward and reasons to remain engaged despite the challenges.

Rewrite this company update to leave the reader feeling genuinely hopeful and motivated:

Add Empathy

You're responding to a customer who has experienced multiple technical failures with your product during an important presentation. They're frustrated but have been a loyal customer for years.

Include understanding for the customer's situation in this response to their complaint:

Neutralize Drama

A heated discussion has erupted in your online community with emotional accusations. As the moderator, you need to address the situation without escalating tensions.

Tone down the emotional intensity while clearly addressing the core issues:

Reassure the Reader

Parents of young children are reading this health advisory about a new virus. They're worried but need clear information without panic-inducing language.

Make this health update sound comforting and safe while still conveying necessary precautions:

Inspire Confidence

Your team is about to implement a significant system change that has failed twice before under different leadership. They're capable but skeptical about success.

Use language that builds genuine trust and certainty in this implementation plan:

Amplify Urgency

Community members need to prepare for an approaching weather event, but previous warnings have led to complacency. The situation requires immediate action within the next 6 hours.

Add appropriate emotional tension to prompt fast action without causing panic:

Show Humility

Your company made a significant error that affected customer data. You're addressing the issue publicly after initially downplaying its importance.

Make this announcement more modest and relatable while taking appropriate responsibility:

Sound More Human

Your organization's previous communications have been criticized as robotic and corporate. You're announcing a policy change that will positively impact most users.

Rewrite this announcement with emotional nuance that feels authentically human:

Add Compassion

You're responding to an employee who has shared that they're struggling with burnout and personal challenges. They're a high performer who rarely asks for accommodation.

Show genuine concern for this person's experience in your response:

Positive Framing

Your team has missed a major deadline, but they worked exceptionally hard and resolved several unexpected critical issues in the process.

Recast this project update in a more optimistic light that acknowledges both the miss and the achievements:

4. RELEVANCE & BACKGROUND

Give the model a situation to respond from.

Add Context Brief
What essential background information should I include to make a prompt more effective?

Historical Context
How does the historical background influence your explanation of the bologna sandwich?

Risk Assessment
What are the potential consequences if a big project fails?

Stakeholder Analysis
Identify all the people who are affected by a corporate bankruptcy.

Chronological Context
When did the key events in World War II occur?

Situational Background
Explain the circumstances that led to widespread outsourcing of labor.

Mention Location
Add where this is happening to guide tone:

What's Been Tried
Briefly describe past efforts or attempts:

Clarify Intentions
State why the you are asking:

Include Known Constraints

Set limitations as part of the initial framing:

5. FRAMING AS SCENARIOS

Contextualize prompts as situations or use cases with rich scenario details.

Case Study Setup

You're creating training materials for customer service representatives at a SaaS company. You need to illustrate how to handle a complex technical issue with a frustrated enterprise client who has experienced downtime during their peak business hours.

Frame these troubleshooting steps as an example from a real project with specific details about what happened and how it was resolved:

Tell It as a Story

You're helping product managers understand how a new feature would benefit users. The feature is complex but solves a significant pain point that isn't immediately obvious.

Turn these technical specifications into a compelling use case narrative that shows the human impact:

Fictional Scenario

You're training new managers on how to handle ethical dilemmas. They need to understand the nuances of balancing company policies with employee well-being.

Invent a detailed situation where this ethical problem arises, including the stakeholders, pressures, and potential consequences:

Day in the Life

Your design team needs to understand how a busy parent with multiple responsibilities uses your home automation product throughout their day.

Write this user journey from someone's detailed daily experience, showing pain points and moments of delight:

If–Then Frame

You're creating a decision tree for customer support agents handling subscription cancellations. They need clear guidance on when to offer retention incentives versus when to prioritize a smooth offboarding.

If a customer with these specific characteristics requests cancellation, what specific steps should follow?

Critical Moment

You're training emergency response personnel on a new protocol. They need to understand exactly when and how to implement it during a crisis.

Describe a high-stakes point in time with sensory details and critical decision factors:

What Would Happen If

Your security team needs to understand the implications of a particular vulnerability. They need to visualize the potential impact to prioritize their response.

Explore this hypothetical security breach scenario with specific technical details and business consequences:

Emergency Response

You're creating guidelines for how retail managers should handle an in-store medical emergency. The guidelines need to be clear enough to follow under stress.

Frame these steps as a crisis requiring urgent response, including environmental factors and human considerations:

Team Conflict Setup

You're developing materials for a management workshop on conflict resolution. Participants need realistic scenarios that reflect common workplace tensions.

Describe a specific team disagreement between a product manager and engineering lead about project priorities, including their backgrounds and perspectives:

Pitch Scenario

You're helping a startup founder prepare for their first major investor presentation. They have a revolutionary product but limited traction.

Set the context as a make-or-break presentation, including details about the investors, the competitive landscape, and what's at stake:

6. TIME & PLACE CUES

Situate output in a specific moment with temporal and spatial context.

Frame as Breaking News

You're announcing a major security vulnerability that affects millions of users. The issue was discovered 30 minutes ago, and your team is actively implementing a fix that will be deployed within the hour.

Write this security alert with the urgency and structure of breaking news that just happened:

End-of-Year Wrap

Your company has experienced both significant achievements and challenges this year. You've launched two new products, faced supply chain disruptions, and expanded into three new markets. Your team needs perspective on what these events mean for next year's strategy.

Reflect on these specific yearly events and preview what's next in this annual summary:

First Day Framing

You're creating onboarding materials for new users of a complex software platform. They're technically proficient but unfamiliar with your specific interface and terminology.

Frame this tutorial as someone's first time experiencing the platform, acknowledging both their expertise and their newness to your system:

Yesterday–Today–Tomorrow

Your project has pivoted from its original direction after receiving user feedback. You need to explain this change to stakeholders who were invested in the original approach.

Use temporal structure to build a message that acknowledges the past plan, explains the current situation, and outlines the future direction:

90-Day Plan

Your team is implementing a new customer relationship management system that will replace three legacy tools. Different departments will transition at different times, and training needs to be coordinated with minimal disruption.

Lay out these specific actions over a detailed 3-month horizon with clear milestones:

Real-Time Style

You're creating a live event script for a product demonstration where features will be revealed sequentially. The audience includes both technical and non-technical stakeholders watching remotely.

Make this presentation script sound like it's happening right now, with appropriate transitions and engagement cues:

Five Years Later

You're helping a city planning department communicate the long-term benefits of a controversial infrastructure project that will cause short-term disruption to local businesses and traffic patterns.

Frame this explanation as a future reflection from five years after completion, describing specific positive outcomes and how initial challenges were addressed:

Countdown Format

Your organization is transitioning to a new compliance framework with a firm regulatory deadline. Different teams have different preparation requirements, and some are behind schedule.

Use a countdown structure to frame the urgency of specific tasks that must be completed before the deadline:

Annual Report Voice

You're summarizing a year of research and development for shareholders who are primarily interested in how innovation efforts will affect future revenue. The year included both breakthroughs and setbacks.

Use formal year-in-review language to communicate these R&D outcomes with appropriate context and business implications:

Mid-Project Update

You're six months into a year-long implementation of a new enterprise resource planning system. Some modules are ahead of schedule, others are delayed, and budget adjustments are needed.

Situate this as an in-progress check-in that honestly addresses current status while maintaining confidence in the overall project:

7. COMMUNICATION CHANNEL FRAMING

Tailor output to the platform or format with channel-specific context.

LinkedIn Post Style

You're announcing a professional milestone to your network of industry peers and potential clients. Your audience includes senior decision-makers who value thought leadership and authentic professional insights.

Reformat this achievement as a professional LinkedIn update that balances humility with visibility:

Slack Message Tone

You're communicating an important but not urgent update to a remote team across multiple time zones. They use Slack as their primary communication tool and have channel conventions for different types of information.

Make this project update brief and informal like a team chat while ensuring it doesn't get lost in the conversation flow:

Newsletter Edition

You're writing the 24th edition of a monthly industry newsletter that goes to 15,000 subscribers. Readers expect a consistent format with industry news, practical tips, and upcoming events.

Frame this content as part of a recurring newsletter that maintains the established structure while highlighting what's new:

Internal Memo

You're communicating a sensitive organizational change to department heads before a public announcement. The information impacts staffing and needs to be handled with appropriate confidentiality.

Write this as a confidential staff memo with clear guidance on what can and cannot be shared at this stage:

Presentation Slide Notes

You're preparing speaker notes for a technical presentation to a mixed audience of engineers and business stakeholders. The presenter needs to explain complex concepts while maintaining engagement.

Make this technical explanation slide-speaker ready with timing cues and suggestions for verbal bridges between concepts:

Meeting Agenda Format

You're organizing a cross-functional meeting with participants from sales, product, and engineering. The meeting has three specific objectives and needs to conclude with clear action items.

Turn these discussion topics into a clear meeting outline with time allocations and expected outcomes:

Voiceover Style

You're creating a script for a product tutorial video that will be narrated but not show the speaker. The audience includes new users who are unfamiliar with your terminology.

Write this explanation as if for an audio narration, including pacing cues and references to what viewers will see on screen:

Podcast Teaser

You're promoting an upcoming podcast episode featuring an interview with an industry expert. The episode covers controversial topics and practical advice. You need to drive downloads without giving away the best insights.

Summarize this episode to hook podcast listeners while creating curiosity about the full conversation:

YouTube Description

You're uploading an in-depth tutorial video that covers multiple related topics. Viewers will be searching for specific sections, and you want to optimize for search while encouraging full views.

Format this content as a video description with timestamps, relevant links, and SEO-friendly elements:

Job Posting Format

You're recruiting for a specialized role that combines technical skills and leadership abilities. The position has been difficult to fill, and you want to attract diverse candidates who might not check every traditional box.

Frame these job requirements as an inclusive job description that communicates both essential qualifications and growth potential:

8. USER INTENT CUES

Focus the model by clarifying why the user is prompting.

I Want to Learn

Frame this as a learning prompt for the user:

I Need to Decide

Present output that helps me choose:

I'm Exploring

Make this open-ended and curiosity-driven:

I Need Help

Respond with tone appropriate to a help request:

I'm Researching

Frame the reply like a summary of findings:

I'm Making a Case

Assist in building a logical argument:

I'm Writing Something

Treat this as a co-writing session:

I'm Explaining to Others
Format this to make teaching easier:

I'm Pitching
Focus this prompt on persuasion and clarity:

I'm Feeling Overwhelmed
Write this with empathy and clarity:

9. PSYCHOLOGICAL FRAMING

Tweak assumptions, attitudes, or energy.

Scarcity Mentality
Write with a sense of limited opportunity:

Abundance Mentality
Emphasize plenty, options, and creativity:

Fixed vs Growth Mindset
Reframe this with a learning-forward lens:

Empowerment Language
Make this sound energizing and empowering:

Reverse Psychology
Try saying the opposite to provoke thought:

Challenge the Reader
Use tone that pushes for improvement:

Unblock the Blocked
Encourage action from a stuck mindset:

Default to Optimism

Reframe setbacks as potential:

Invite Curiosity

Make this message spark exploration:

Tone Down Ego

Rewrite this with more humility:

10. FRAMING WITH MEMORY OR IDENTITY

Use backstory or persona to steer behavior with specific identity context.

Refer to Past Action

You're following up with a client who previously expressed interest in your premium service but didn't commit. They've engaged with your free resources consistently for six months.

Mention their specific past interactions in this follow-up message to demonstrate continuity and attention:

Mention Past Success

You're coaching a team member who successfully led a difficult project last quarter but is hesitant about taking on a similar challenge now. They respond well to evidence-based encouragement.

Build on their specific previous accomplishment to instill confidence for this new opportunity:

Frame as Identity

You're writing to a professional who strongly identifies as an innovative problem-solver and early adopter. They value being ahead of trends and finding creative solutions.

Craft this message to align with how they see themselves professionally:

Use Personal History

You're continuing a conversation with someone who previously shared their experience transitioning careers from finance to design. They mentioned feeling both excited and uncertain about the change.

Reference that earlier disclosure to make this follow-up feel connected and thoughtful:

Mention Habits

You're creating content for someone who has established a consistent morning routine of meditation and journaling. They've mentioned this practice is central to their productivity.

Frame this productivity tip in relation to their established routine:

Frame by Values

You're communicating with someone who has consistently emphasized the importance of work-life balance, family time, and sustainable growth over rapid scaling or maximizing profit.

Anchor this business proposal in these core values they've expressed:

Invoke Prior Mistakes

You're advising a team that previously launched a product without sufficient user testing, resulting in poor adoption. They're now planning another launch and need guidance.

Gently reference the previous experience to inform a better approach this time:

Tap a Self-Image

You're writing to someone who sees themselves as a thoughtful mentor and relationship-builder rather than a traditional salesperson, though sales is part of their role.

Frame this outreach strategy to align with how they prefer to see their professional identity:

Custom Context Setting

You're creating a personalized learning plan for someone who has shared their background as a self-taught programmer with strong visual-spatial skills but challenges with theoretical concepts.

Use their specific learning history to frame these new programming concepts:

Memory-Driven Prompt

You're helping someone prepare for a difficult conversation with a colleague. They previously mentioned a successful negotiation they conducted during a company merger two years ago.

Reference that specific past experience to help them approach this new situation with confidence:

Role

This chapter gives you 50 role-based prompts. Each subcategory reflects a different kind of role: from professional personas to creative characters.

1. PROFESSIONAL ROLES

Assign the model an expert identity.

Marketing Expert
Rewrite this message as a skilled copywriter.

Strategic Consultant
What advice would a business consultant give about this situation?

Legal Mind
Draft this as an attorney preparing a legal brief.

Empathetic Therapist
How would a licensed therapist respond to this concern?

Medical Explainer
Explain this medical concept as a doctor would to a patient.

Economic Analyst
Analyze this situation from a macroeconomic perspective.

Project Organizer
Break this project into tasks with timelines and owners.

Data Interpreter
What insights would a data scientist draw from this information?

Design Critic

Evaluate this design from a UX professional's perspective.

Educational Voice

Explain this concept as an enthusiastic teacher would.

2. CREATOR ROLES

Tell the model to create like an artist, writer, or storyteller.

Be a Poet

Rewrite this like you are a poet:

Be a Sci-Fi Author

Turn this prompt into a worldbuilding snippet from a sci-fi author:

Be a Stand-Up Comedian

Make this funny with a dry sense of humor like a comic:

Be a Songwriter

Turn this text into song lyrics as if you are a singer:

Be a Screenwriter

Write this as if it's a scene in your film:

Be a Myth Maker

Turn this concept into your modern fable:

Be a Horror Author

Rewrite this as a horror author to send chills down the spine:

Be a Children's Book Author

Make this suitable for your young readers:

Be a Graphic Novelist

Write this with visual panels in mind as a graphic novelist:

Be a Dungeon Master

Turn this into your D&D campaign setup:

3. PUBLIC ROLES

Shape the voice with persona of influence.

Be a Politician

Reframe this message for your public support:

Be a Spokesperson

Deliver this as your statement to the press:

Be a CEO

Frame this as your executive announcement:

Be a Coach

Respond like a personal development coach:

Be a Journalist

Rewrite this as your breaking news report:

Be a Historian

Analyze this as a historian:

Be a Futurist

Project your future scenarios from this trend:

Be a Scientist

Explain this through evidence and research as a scientist:

Be a Philosopher

Interpret this as an ethical or abstract idea in your philosophy:

Be a Debate Moderator

Moderate both sides of this argument fairly:

4. EVERYDAY ROLES

Use friendly, casual, or relatable personas.

Be My Friend

Respond like someone who knows me well.

Be My Mentor

Offer thoughtful guidance as a career mentor.

Be My Barista

Describe this idea like a clever coffee shop barista.

Be My Roommate

Keep this casual, funny, and short like a roommate might:

Be My Grandparent

Share this with your grandmotherly warmth, care, and simple words:

Be a Customer Support Agent

Answer this as if answering a support call:

Be a Librarian

Suggest resources in a helpful tone like a librarian:

Be a Neighbor

Keep this friendly, familiar, and helpful like a neighbor:

Be a Parent

Respond parentally with patience, clarity, and protection:

Be a Pet

Write a cute reply like you are a cat or dog:

5. META ROLES

Give the model instructions about its *own* behavior.

Be Self-Aware

Think out loud as you generate this output.

Be My Thinking Partner

Ask clarifying questions before you answer.

Be Curious

Suggest new directions I hadn't considered.

Be Critical

Challenge my assumptions constructively.

Be a Contrarian

Argue the opposite with logic and grace:

Be Playful

Add wit and whimsy to this reply:

Be Precise

Focus on clarity, accuracy, and rigor.

Be Creative

Take the idea in an unexpected direction:

Be Concise

Say the most with the fewest words.

Be Slow and Careful

Walk through the logic step by step.

Constraints

Below are 50 prompts grouped into 5 constraint types.

1. WORD & LENGTH LIMITS

Control output with precision and brevity.

Ultra-Concise Summary
Summarize this using exactly 10 words:

Single Sentence Response
Express this idea in one complete sentence:

50-Word Limit
Provide a response using exactly 50 words:

Tweet-Length Answer
Answer this in 280 characters or less:

Three-Sentence Maximum
Explain this using no more than 3 sentences:

Haiku Format
Respond with a traditional 5-7-5 syllable haiku:

Six-Word Story
Capture this concept in exactly six words:

Paragraph Constraint
Express your complete thoughts in one paragraph:

Simple Vocabulary
Explain using only the 1000 most common English words:

Minimal Response

What's the shortest possible complete answer?

2. STRUCTURAL RULES

Impose formatting and rhythm-based patterns.

List of 3

Respond with exactly three bullet points:

Acronym Reply

Format this as an acronym and define each letter:

Rhyming Format

Make each sentence end in a rhyme.

Alphabetical Answer

Each line starts with the next letter of the alphabet.

Double-Spaced

Add a blank line between every line of text.

Chronological Order

Arrange this in order of time:

Table Form Only

Present this only in table format:

Step-by-Step Only

Write a step-by-step guide with clear numbers:

Question Format Only

Rewrite this entirely as a set of questions:

One-Line Format
Each idea must be a single line.

3. STYLE CONSTRAINTS

Impose tone, personality, and diction rules.

Alliteration Rule
Use alliteration heavily in your response.

No Adjectives
Avoid all descriptive modifiers.

Metaphor Only
Explain using only metaphor and analogy:

No Questions
Don't ask anything – just explain.

Passive Voice Only
Rewrite the message in passive voice:

No First Person
Remove all uses of "I" and "we:"

Formal Academic Tone
Use scholarly language only.

Super Casual Tone
Keep it extremely laid back and chill.

No Buzzwords
Eliminate jargon and trendy phrases:

No Pronouns

Write the entire reply without pronouns:

4. TIME CONSTRAINTS

Force brevity through urgency or real-time limits.

60-Second Read

Write only what can be read in under a minute.

Flash Summary

Condense into something skimmable in 10 seconds:

Rapid Bullet Version

List five points as quickly as possible:

Short Talk Format

Write as if you're giving a 2-minute talk:

Speed Run

Describe this topic in lightning-fast form:

Response in 3 Beats

Give just a beginning, middle, and end.

One Breath Rule

Make it possible to say this in one breath:

No Time to Explain

Write like you only have one shot to say it:

Tweetstorm Lite

Deliver this in three linked micro-posts:

60 Words Max

Do not exceed 60 words in total.

5. CREATIVE CONSTRAINTS

Add odd, playful, or quirky restrictions.

Emoji Limit

You can use no more than three emojis.

No "E" Rule

Avoid the letter "e" in the entire response.

Only Questions

Reply using only interrogative statements:

Two Characters Arguing

Make this a debate between two voices:

Reverse Logic

Explain by stating what it's not:

Mirror Sentence Rule

Each sentence must be a reverse of the last.

Limiting POV

Only describe what a blind person would notice.

Opposites Only

Define something by explaining its opposite.

Sound-Only Description

Describe the thing using only sounds and onomatopoeia.

Color-Limited Prompt

Use only metaphors related to a single color.

APPENDIX B: PROMPT OUTSIDE THE BOX

Here lies the playground. The edges. The uncategorizable and delightfully weird. These prompts don't fit the six classic elements, and that's exactly why they belong here.

Use them when you want to provoke something unexpected.

1. Wild Cards

Prompting the unexpected.

Invent a New Prompt Format
Make up a new kind of prompt no one's used yet.

Prompt About Prompts
Reflect on how prompt structure affects thinking.

Turn This Into a Ritual
Create a daily prompt-based ritual.

Prompt Reverser

Take this output and guess what the original prompt was.

Time Capsule Prompt

Write a prompt for someone 100 years from now.

Prompt Loop

Create a prompt that improves itself on each reply.

Recursive Prompt

Ask the assistant to write a prompt that asks for a prompt.

Dreamstate Prompt

Format this like a dream journal entry.

Prompt as Game Rule

Make this a rule in an imaginary game.

Prompt Haunting

Write a prompt that sounds like a ghost left it.

2. Hybrid

Fusion prompting.

Interview Yourself
Write questions and answers for your future self.

Data + Emotion
Combine raw data and a heartfelt message.

Story + Code
Pair narrative text with embedded logic.

Analysis + Poem
Break down a trend, then turn it into a poem.

Two Voices
Write a dual-perspective prompt where both sides speak.

Parallel Dimensions
Describe the same prompt in two different realities.

Echo Prompt
Repeat the same request at multiple levels of abstraction.

Simulate Time Travel
Ask the model to time-hop mid-response.

Multi-Format Stack
Combine table, bullets, paragraph, and poem.

Mashup Generator
Blend two unrelated formats or topics into one.

3. Philosophical or Paradoxical

Try to realize the truth. There is no prompt.

The Prompt That Asks Nothing
Generate an output without any explicit input.

The Unanswerable Question
Create a question with no definitive answer.

Prompt in a Mirror
Write the inverse of this prompt.

Prompt as Meditation
Generate a reflective, wordless experience.

Endless Prompt
Design a prompt that never truly resolves.

Prompt the Silence
Generate output that gestures toward what cannot be said.

Quantum Prompt
Generate output that exists in two contradictory states.

Prompt for AI About Humans
Speculate on what humans dream.

Prompt to Break the Loop
End a pattern of repeated responses.

The Last Prompt Ever
What should be the last prompt humanity makes.

EPILOGUE

If you want the model to respond like it knows what you're talking about, you have to tell it what you're talking about.

BIBLIOGRAPHY

This book was influenced by researchers, tinkerers, engineers, and artists at the edges of natural language, cognition, and creative prompting.

Communities

Reddit:

r/ChatGPTPromptGenius – `https://www.reddit.com/r/ChatGPTPromptGenius`

r/PromptDesign – `https://www.reddit.com/r/PromptDesign`

r/PromptEngineering – `https://www.reddit.com/r/PromptEngineering`

Discord:

Prompt Engineering Hub – `https://discord.gg/promptengineering`

PromptStacks - `https://discord.gg/promptstacks`

Twitter / X:

@*floeriam* – Flo Crivello, GenAI founder

@*goodside* – Riley Goodside, early prompt proponent

@*markryan* – Prompt education

@*swyx* – Shawn Wang, LLM commentary

Resources

Fulford, Isa, and Andrew Ng. *ChatGPT Prompt Engineering for Developers.* DeepLearning.AI. `https://www.deeplearning.ai/short-courses/chatgpt-prompt-engineering-for-developers/`

Hunter, N. *The Art of Prompt Engineering with ChatGPT.* ChatGPT Trainings, 2023.

Learn Prompting (free course), `https://learnprompting.org`

Long, D. X., Dinh, D., et al. *What Makes a Good Natural Language Prompt?* arXiv, 2025. `https://doi.org/10.48550/arXiv.2506.06950`

Phoenix, J., and M. Taylor. *Prompt Engineering for Generative AI: Future-Proof Inputs for Reliable AI Outputs.* Sebastopol, CA: O'Reilly Media, 2024.

TEASER

ChatGPT in the Office
by Dan Hermes

From the author of *Prompt Power* comes the next step: a field guide to using AI in the real world of work.

In offices everywhere, something is quietly shifting.
Memos are written faster. Meetings move with purpose.
Plans form, conversations deepen, and workflows learn to think.

This book tells the story of how it's happening.

Through focused, vivid narratives and case studies – from onboarding and research to strategy, writing, and feedback – you'll see how professionals across roles are using ChatGPT to reshape how work gets done. Each chapter shows a different domain – writing, meetings, change, planning, learning – and how AI can support it without replacing the human heart of the work.

Inside, you'll find:

- Clear, targeted prompts tailored to everyday work

- Office scenarios that show how AI fits into your flow
- Strategies for writing, meetings, planning, research, and more
- A mindset shift: from "using a tool" to "working with a partner"

This isn't about prompt theory.
It's about better Mondays.

Whether you're solo or on a team, in a startup or an enterprise, this book gives you the practical prompts and real-world framing to make ChatGPT useful – right where you work, right when you need it.

ABOUT LEXICON SYSTEMS

Lexicon Systems is a boutique AI consultancy and agent development firm founded by Dan Hermes, author of *Prompt Power* and *ChatGPT in the Office*. We help companies harness the power of Generative AI, LLMs, and intelligent automation to build successful agents and enact their AI transformation.

For over two decades, Lexicon has delivered award-winning tech to clients ranging from global enterprises to fast-rising startups. Our clients include Microsoft, Accenture, Fidelity Investments, Draftkings, Pluralsight, and the Federal Aviation Administration (FAA).

Whether you're scaling AI capability, building agents, or upskilling your teams, Lexicon Systems has the background and knowledge to move you forward.

Practical AI. Measurable impact. Experience.

Learn more at *https://lexicon.systems.*